T0332455

Global Health Challenges

This book is an up-to-date reference on some major global health issues and the role of nutrition in their prevention and management. The book covers undernutrition, degenerative diseases, mental health disorders, and COVID-19 and reviews feeding and eating disorders like anorexia, bulimia, binge-eating, and delineates the risk factors and management. The book also addresses the gaps in tackling these health problems and proposes comprehensive models and frameworks to manage them.

Key features:

- Explores practical solutions and management of looming health issues in terms of diet and nutrition
- Reviews health threats like obesity, diabetes, hypertension, and COVID-19
- Includes a section on feeding and eating disorders and sustainable models to manage them
- Covers mental health issues like depression and dementia
- Discusses conditions like undernutrition, hidden hunger, and cardiovascular diseases

The book is meant for health professionals, nutritionists, and policymakers. It is also useful for post-graduates in public health and nutrition.

Global Health Challenges

Nutrition and Management

Sarita Srivastava, Anju Bisht, Avula Laxmaiah,
and Seetha Anitha

CRC Press
Taylor & Francis Group
Boca Raton London New York

CRC Press is an imprint of the
Taylor & Francis Group, an **informa** business

Cover image courtesy of Authors

First edition published 2024
by CRC Press
2385 NW Executive Center Drive, Suite 320, Boca Raton FL 33431

and by CRC Press
4 Park Square, Milton Park, Abingdon, Oxon, OX14 4RN

CRC Press is an imprint of Taylor & Francis Group, LLC

ISBN: 9781032378466 (hbk)
ISBN: 9781032406299 (pbk)
ISBN: 9781003354024 (ebk)

DOI: 10.1201/9781003354024

Typeset in Times
by Deanta Global Publishing Services, Chennai, India

Contents

Foreword

राष्ट्रीय आयुर्विज्ञान अकादमी (भारत)
अन्सारी नगर, महात्मा गांधी मार्ग, नई दिल्ली-110029
दूरभाष : 26588718
        26589289

NATIONAL ACADEMY OF MEDICAL SCIENCES (INDIA)
Ansari Nagar, Mahatma Gandhi Marg, New Delhi - 110029
Telephone : 011-26588718   Email: nams_aca@yahoo.com
              011-26589289   umeshkapil@gmail.com
Mobile   : 9810609340   Website : www.nams-india.in

Dr. Umesh Kapil
MD, FAMS,
Secretary

At the current time, the double burden of communicable and non-communicable diseases, coupled with an unprecedented pandemic like COVID-19, has engulfed the health of a large segment of the vulnerable population. Global health issues need to be addressed, particularly by strengthening preventive measures.

This book talks about the spectrum of current health challenges ranging from physiological disorders to mental health issues. Chapters in the book focus in detail on health issues, viz., COVID-19, undernutrition and hunger, obesity, diabetes, cardiovascular diseases, cancer, feeding and eating disorders, dementia, and depression with special emphasis on their comprehensive management approach.

The book is essential reading for health professionals, students, and all those for whom health is of prime importance. I appreciate the efforts put in by the authors.

(Umesh Kapil)

Message

सतपाल महाराज
मंत्री

लोक निर्माण विभाग, पर्यटन, सिंचाई, लघु सिंचाई ग्रामीण निर्माण
पंचायती राज, जलागम प्रबन्धन, संस्कृति, धर्मस्व
भारत-नेपाल उत्तराखण्ड नदी परियोजनाएं
उत्तराखण्ड सरकार
Public Works Department, Tourism, Irrigation, Minor Irrigation,
Rural Works Department, Panchayati Raj, Watershed Management, Culture
Spiritual Department, Indian-Nepal-Uttarakhand River Projects
Government of Uttarakhand

I am glad to know that a book entitled *Global Health Challenges: Nutrition and Management* is going to be published for the awareness of health issues. Health, encompassing both physical and mental wellbeing, is an inherent right for every individual. However, regrettably, in our fast-paced world, health challenges are also evolving simultaneously. The enduring presence of well-known health issues like undernutrition, alongside the emergence of new challenges such as COVID-19, presents a severe threat to overall wellbeing.

Global Health Challenges: Nutrition and Management addresses key health issues, including COVID-19, undernutrition, childhood obesity, diabetes, cancer, cardiovascular diseases, feeding and eating disorders, depression, and neurodegenerative disorders. The book aims to explore the causes, consequences, and management of these health challenges. This book is also unique in its integration of both physical and mental health challenges in one source, enhancing the depth and breadth of knowledge.

I hope that this book will offer an effective learning experience for students in the field of health and nutrition. Furthermore, it can serve as a valuable reference for healthcare professionals, policy-makers, educators, and the general public concerned about their health.

I appreciate this effort of the authors and extend my best wishes for the publication of the book.

(Satpal Maharaj)

Preface

The well-known saying 'health is wealth' carries profound significance. Indeed, health surpasses mere material wealth. It plays a pivotal role in leading a wholesome life and realising one's full potential. By addressing global health concerns through prevention and control, we can prevent the suffering and loss of lives of millions, while also minimising economic burdens. Historical records show that ancient Indian sages adhered to a disciplined regimen of yoga, meditation, and a clean, unadulterated diet in a pure environment, potentially resulting in extended lifespans with robust physical and mental wellbeing. These insights and observations motivated a group of nutritionists and medical professionals to collaboratively tackle global health challenges.

This book is written in clear, accessible language, ensuring easy comprehension for readers. It covers a range of topics including COVID-19 infection, malnutrition and hidden hunger, childhood obesity, diabetes, cardiovascular diseases, cancer, feeding and eating disorders, depression, and neurocognitive disorders (such as dementia). Each chapter emphasises a comprehensive approach to managing these health issues.

This book will prove valuable not only to healthcare professionals but also will serve as a practical guide in preventing and controlling some of the most prevalent diseases seen worldwide today.

Authors
Sarita Srivastava
Anju Bisht
Avula Laxmaiah
Seetha Anitha

Authors

Sarita Srivastava is Director of an NGO, Suvernada Foundation, Rudrapur, Uttarakhand, India, and former Professor and Head of the Department of Foods and Nutrition at GB Pant University of Agriculture & Technology, Pantnagar, Uttarakhand, India. Sarita received an MSc in Foods and Nutrition and a PhD in Biochemistry from GB Pant University of Agriculture & Technology and a Post Doctorate from Botanisches Institut, der TU Braunschweig, Germany (1988–1989). She was the recipient of an FAO fellowship and worked under a UNDP project at Michigan State University, USA, in 1993. She has been a subject editor of *Pantnagar Journal of Research*. She has 39 years of teaching experience and has worked extensively on the nutritional quality of millets and product development for diversified uses, including food products for diabetics and celiac disease patients. In this area, she has guided 30 MSc and 16 PhD students. She is the editor of a book, *Small Millet Grains—The Superfoods in Human Diet* and has 90 research papers and 42 popular articles to her credit. She has also filed six patents. Dr Srivastava has received several awards for excellence and appreciation of her professional work. She received the Best Teacher Award from GB Pant University of Agriculture & Technology in 2002. She has also served as an expert in various national forums and international conferences.

Anju Bisht is Professor in the Department of Home Science at MB Government PG College in Haldwani, Uttarakhand, India. She has more than 18 years of teaching and research experience. She completed her MSc (Foods and Nutrition) and PhD (Human Nutrition) at GB Pant University of Agriculture & Technology. Her research topics were the development of weaning food and functional food for diabetics from small millets. She has published 65 research papers and chapters in edited books, participated in a variety of seminars, and has served as a subject expert. She is a member of professional bodies such as Nutrition Society of India (NSI) and Home Science Association of India (HSAI).

Avula Laxmaiah is a Former Directorgrade Scientist with the ICMR-National Institute of Nutrition, Hyderabad, India, and has an academic background in Medicine and Public Health (MPH), with hands-on professional experience in Public Health Nutrition. He has carried out several large-scale epidemiological studies in Public Health Nutrition and has helped the framing of many nutrition intervention programmes in India. He has published more than 150 papers in national and international peer-reviewed journals like *The Lancet, The Lancet Global Health, Nature, eLife, Obesity Research, BMJ, PHN, Maternal and Child Nutrition, IJMR*, etc.

As a lead public health nutrition scientist, he has developed cereal-pulse-milk-based ready-to-eat and take home ration (THR) supplementary food (Balamrutham) for children aged 6–35 months. Balamrutham is now being distributed to children aged 6–36 months in the Integrated Child Development Services (ICDS) centres in the states of Telangana and Andhra Pradesh. He is instrumental in creating various intervention models for risk reduction of the triple burden of diseases. He was the National Secretary for the Nutrition Society of India (NSI) from April 2020 to March 2024. He has received many national and international awards for his significant work. He is the fellow of National Academy of Science of India (FNASI).

Seetha Anitha started her career as a post-doctoral research fellow in the field of food safety and nutrition at International Crops Research Institute for the Semi-Arid Tropics (ICRISAT), Malawi. She believes in integrating agriculture-based nutrition interventions with health and hygiene through a value chain approach. She completed her PhD in food science and has 13+ years of working experience in Malawi, Tanzania, Myanmar, and India in the field of public health nutrition. Her major achievements are, firstly, introducing millets and pigeonpea in school feeding programmes as a pilot study in Karnataka state in India and in Tanzania. Based on her study, the government of India suggested including millets in all the state's mid-day meal programmes. Secondly, she developed two diagnostic kits to assay aflatoxin biomarkers in human blood and in crops to mitigate aflatoxin contamination and exposure. Finally, she led a large team of scientists to conduct a systematic review and meta-analysis to generate science-based evidence on millets' potential in managing diabetes, lipid profile, iron deficiency anaemia, calcium deficiency, and growth. This series of papers revealed the nutritional and health benefits of millets and was published recently, providing important evidence for millets in the International Year of Millets, 2023. She believes that the culturally sensitive and participatory approach is key to introducing indigenous underutilised crop products into the food basket, bringing behaviour change, improving nutritional status, and scaling, especially among the young who are the future leaders.

1 Global Health: Challenges and Management

1.1 INTRODUCTION

Health is one of the primary needs and rights of every individual across the globe. In the present time, every nation is facing the impact of poor health. The causes and types of illnesses may vary from country to country, but the ultimate impact is on health. The high morbidity, mortality, and Disability Adjusted Life Years (DALYs) index hampers the productivity of the individual and nation as a whole. It therefore becomes important to address health in a global context.

Global health is an area for study, research, and practice that places a priority on improving health and achieving equity in health for all people worldwide (Koplan et al. 2009). Beaglehole and Bonita (2010) proposed the definition of global health as collaborative, transnational research and action for promoting health for all.

Considering the above definitions, we propose that the aim of global health should be to identify and quantify health issues; develop and implement curative measures; and disseminate preventive measures. These goals need to be achieved with the active involvement of all nations focusing on health.

1.2 CHALLENGES OF GLOBAL HEALTH

A lot of communicable and non-communicable diseases have emerged in the past, of which many have been controlled with robust healthcare practices. However, some, along with new diseases, are still surfacing. These illnesses have created havoc for mankind. They are responsible for an increase in premature deaths and have directly impacted the productive years of individuals and their caregivers.

The major contemporary health challenges are recognised as malnutrition, certain degenerative diseases and infections, and mental health disorders. This book attempts to discuss the burning health issues prevailing at the current time. COVID-19, undernutrition and hidden hunger, obesity, diabetes, Cardiovascular diseases, cancer, feeding and eating disorders, depression, and dementia (severe neurocognitive disorder) are the major health challenges affecting various nations and therefore these issues are discussed in detail. These health issues have affected one or another segment of the population, irrespective of age, gender, socioeconomic status, and country of residence. Hence, it becomes important to understand why some diseases are uncontrolled despite the continuous understanding of their causes and advancements in their diagnosis and treatment techniques. Moreover, the sudden emergence of an unprecedented pandemic like COVID-19 calls for an evaluation of the preparedness of countries in tackling such health issues in the future.

1.3 CAUSES OF GLOBAL HEALTH ISSUES

As we all know, understanding the aetiology and risk factors predisposing any health disorder are crucial for the management of the disorder. Today, if we have to work on the eradication of any health disorder, we need to understand its causes not only in terms of the direct environment but also the indirect environment.

DOI: 10.1201/9781003354024-1

The factors affecting health may be broadly classified into three categories. The first category is the primary environment, which refers to individual factors like age, sex, personal vulnerabilities, genetic makeup, and lifestyle factors. The secondary environment comprises the factors which form the individual's immediate environment and directly influence health. These factors include socioeconomic status, housing conditions, poverty, education, employment, safe water, safe food, hygiene and sanitation, home and work environment, neighbourhood conditions, social ties, pollution, air quality, and healthcare facilities. The tertiary environment encompasses climate change, war, conflicts, world trade policies, political machinery, forced migration, natural calamities, population explosion, inadequate and inappropriate food production and processing, which indirectly affect health in the long run. Tertiary factors may influence the secondary factors and consequently the health of an individual.

1.4   MANAGEMENT OF GLOBAL HEALTH CHALLENGES

Managing a disease has two broad dimensions: prevention and treatment. The modifiable risk factors should be the target for the control and prevention of a health disorder. It has been seen that in the past few decades, with the advancement in technology and urbanisation, a huge transition in lifestyle has occurred. Unhealthy diet, sedentary lifestyle, tobacco use, and alcohol misuse have given rise to lifestyle diseases such as obesity, cancer, diabetes, and heart diseases. Urbanisation has exposed people to various environmental pollutants, unhygienic living conditions, and unsafe food and water, leading to an increased risk of infectious and non-infectious diseases. Rapid urbanisation along with an increasing population has put extreme pressure on the healthcare system, impeding its efficiency in providing health services to all those in need. Food insecurity and nutrition insecurity in developing nations have encouraged undernutrition and hidden hunger. The vulnerable segments, such as children under age 5, pregnant and lactating mothers, and adolescent girls, are the main victims of food and nutrition insecurity.

The priority area of treatment is presently focused on modern (allopathic) medicine in treating diseases. Despite the high cost, side-effects, and demand–supply gap in terms of infrastructure and trained personnel, the importance of modern medicine cannot be ignored. However, the holistic treatment of any disease needs to incorporate the traditional medicine system along with complementary and alternative medicine. The traditional Indian medicine system – includes Ayurveda, Yoga & Naturopathy, Unani, Siddha, or the traditional Chinese medicine system which make use of certain herbal preparations and other traditional techniques to treat disease should be revived. All the medicinal systems should be intertwined and practised globally for tackling health challenges.

Diet is the cornerstone of managing all health issues. The common saying 'you are what you eat' denotes the irrefutable connection between food and health. Food has curative as well as preventive properties, and on the other hand, unhealthy food is a risk factor for many disorders. Unhealthy food predisposes individuals to disease conditions such as cancer, obesity, diabetes, heart diseases, infections, and mental illnesses. Immunity-boosting food, antioxidant-rich food, low glycaemic index food, and vegetarian food help to heal the body. The treasure of traditional knowledge should be documented and put into action. Spices and herbs with therapeutic properties should be made an integral part of every cuisine. Food rich in saturated fats, sugar, salt, and low in fibre should be avoided. Switching back to a traditional food system is highly recommended for human and environmental health across the globe.

Industrialisation in the 21st century has made life easier but at the cost of health. Automation has paved the way for a sedentary lifestyle, a major threat to health throughout the globe. Physical inactivity puts us at risk of physical and mental disorders. Physical activity has health benefits and reduces stress. Stress is the key element in promoting psycho-physiological disorders. Coping with stress is the best means of preventing various ailments. Physical activity of any kind, such as walking, running, weight training, aerobics, yoga, and meditation should be adopted for maintaining physical and mental health. Investing in physical activity should be adopted for lifelong health benefits. It is one of the best preventive modalities.

Certain other therapies like stimulatory therapies and psychotherapies are also gaining importance in managing various mental disorders like depression, dementia, and eating disorders. These therapies need to be promoted and made affordable and accessible.

The management and prevention of health issues need to be done at two levels. The primary level is at the individual level. The affected person should adopt a healthy lifestyle and seek professional help for treating and preventing health challenges. The secondary level is focused on the willpower of governments and policymakers. Strengthening the healthcare system so as to make it accessible, affordable and available for all and improving socioeconomic status, such as providing education, employment, and proper living conditions, should be a priority. Generating awareness should be the main agenda. Imposing strict regulations and enforcing laws related to health hazards and promoting research in the fields of medicine, immunisation, etc., are needed.

1.5 CONCLUSIONS

WHO defined 'Health is a state of complete physical, mental and social well-being and not merely the absence of disease or infirmity'. Health is physical, mental, and social wellbeing. It is one of the primary needs of every individual and every nation. Besides the noteworthy advances in maintaining health and fighting illnesses, the world is still besieged with many health challenges. Currently, COVID-19, undernutrition and hidden hunger, obesity, diabetes, cardiovascular disease, cancer, eating disorders, depression, and dementia (severe neurocognitive disorder) are the most challenging health problems that are leading to premature death and disability. These health issues should be the centre of attention across the globe. The aetiologies for many are known, yet these diseases are on the rise, and therefore their cure and prevention need to be thought about from a newer perspective. A multidisciplinary mixed approach should be customised. The shared responsibility of individuals, medical professionals, nutritionists, psychologists, policymakers, and governments can address these global health issues. Exchanging constructive relevant research and healthcare practices across borders can be helpful. Every nation has to take responsibility on a large scale. Global warming, climate change, pollution, wars, conflicts, forced migration, droughts, floods, and famines should be tackled as a global issue. Thus, it becomes the joint responsibility of every individual, concerned authority, and government to manage the health challenges for a healthier and happier world.

REFERENCES

Beaglehole, R., and R. Bonita. 2010. What is global health? *Global Health Action* 3. https://doi.org/10.3402/gha.v3i0.5142

Koplan, J.P., T.C. Bond, M.H. Merson, et al. 2009. Towards a common definition of global health. *Lancet* 373: 1993–1995.

2 COVID-19 Infection

2.1 INTRODUCTION

Infection can be understood as the invasion and multiplication of disease-causing agents like viruses, bacteria, fungi, protozoa, and helminthes in the host cells, causing damage to the normal functioning of host cells or production of toxins and damaging cells, thereby causing disease. Normally, the disease is caused under suppressed immunity conditions (Drexler 2010).

The mode of transmission of infectious agents to a susceptible host can be categorised as direct and indirect contact. In direct transmission, the infectious agent from a reservoir reaches the host through direct physical contact, direct droplet transmission, direct exposure to an infectious agent in the environment, animal bites, and transplacental and perinatal transmission, whereas in indirect transmission, the infectious agent is transferred indirectly from a reservoir or infected host to a susceptible host. It can be through a biological vector or intermediate host, a mechanical vector or vehicle, and airborne (van Seventer and Hochberg 2017).

The disturbing relationship between humans and microorganisms due to climate change, encroachment of the human population into forests, and human displacement because of socioeconomic pressure has brought the human population into contact with new pathogens. War, famine, and floods are some controllable and uncontrollable factors that aggravate the chances of infections as they lead to human displacement, imbalancing the ecosystems. Furthermore, the export and import of food across national and international borders may contaminate food in the supply chain, thereby increasing the chances of infection. Poverty is another factor that is accountable for the increase in episodes of infection. It is directly related to malnutrition and infection. Poverty forces the poor to migrate and settle in congested areas with poor access to clean water and sanitary conditions, adding to the incidence of infection (Drexler 2010). The major risk factors for infectious diseases are malnutrition, poor water supply, poor sanitation, and unhygienic conditions, unsafe sex, use of tobacco and alcohol, drug abuse, occupational hazards, sedentary lifestyle, air pollution, of which malnutrition, poor water supply, poor sanitation, and unhygienic conditions are major risk factors for infectious diseases in children (Michaud 2009).

Infectious diseases have taken a heavy toll on life across the world. Exploring the trends in the prevalence of infectious diseases globally, it can be seen that in 1990, 17.2 million deaths were due to infectious and parasitic diseases, respiratory infections, and maternal and perinatal disorders (Murray and Lopez 1997) and in 2001, 14.7 million deaths were due to infectious diseases, chiefly in developing regions, i.e., 6.8 million in sub-Saharan Africa and 4.4 million in South Asia (Michaud 2009). These data support the fact that with the understanding of techniques to prevent, diagnose, and treat infectious diseases, a decline in the percentage of infected people has occurred; however, the overall number of outbreaks has increased (Smith et al. 2014).

Despite the progress in controlling and treating infectious diseases, the resurgence of old infectious diseases and the emergence of novel infectious diseases like COVID-19 always remain a threat to mankind. In this chapter, the COVID-19 pandemic, which has surfaced as the biggest human health crisis in the present time, is discussed.

2.2 COVID-19 INFECTION

COVID-19 (Corona Virus Disease-19) is a highly contagious infectious disease affecting the respiratory system. The first case of COVID-19 was reported in China in December 2019, from where

DOI: 10.1201/9781003354024-2

it spread rapidly across the world, wreaking havoc on lives and livelihoods and hence it being declared as a global pandemic by the World Health Organization (WHO).

COVID-19 is caused by the SARS-CoV-2 virus. SARS-CoV-2 is a single-stranded RNA-enveloped virus. The surface of SARS-CoV-2 is covered by glycosylated S protein, which binds to host cells and facilitates entry into the cell (Huang et al. 2020), primarily respiratory epithelium. Once the virus invades the respiratory system, it develops disease in two phases. In the initial phase, virus replication occurs, which damages tissues, and in the later phase, the infected host cell triggers an immune response. Under severe condition, the overactivation of the immune system results in the release of high levels of cytokines, causing inflammation and consequently fibrosis. Besides the respiratory system, SARS-CoV-2 can affect the gastrointestinal tract, cardiovascular system, kidneys, and central nervous system, leading to organ dysfunction (Cascella et al. 2021).

2.3 MODE OF TRANSMISSION

The probable modes of transmission of the SARS-CoV-2 virus from infected to non-infected individuals are through respiratory droplets, airborne transmission, fomite transmission, faecal-oral transmission, and vertical transmission from mother to neonate as shown in Figure 2.1 (Cascella et al. 2021). When a person is in close contact (within 1 metre) with an infected person, the virus can be transmitted by coughing, sneezing, singing, or talking loudly in the form of respiratory droplets or aerosols. The virus can enter a healthy person through their nose, mouth, and eyes. Airborne transmission refers to when the infectious aerosol remains suspended in the air for a long period of time and a susceptible person inhales the aerosol, leading to infection. The chance of airborne transmission is high in crowded and poorly ventilated areas. Fomite transmission is an indirect type of transmission. In fomite transmission, the susceptible person comes into contact with a contaminated surface or object (fomite) and becomes infected. Faecal-oral transmission and transmission from mother to infant through breastmilk require more substantial research for conclusive results (WHO 2020a).

2.4 SUSCEPTIBLE TARGETS

Although every individual is equally prone to infection by the virus, it was seen that the geriatric population, males, smokers, and patients with comorbidities like diabetes, obesity, kidney diseases, lung diseases, and cancer were at a greater risk of developing severe infection and higher mortality.

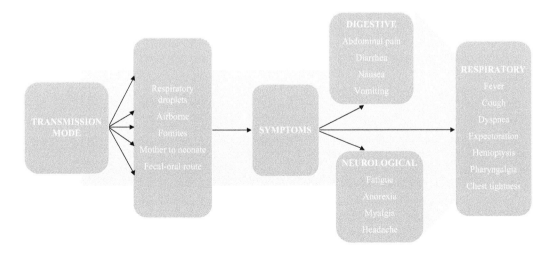

FIGURE 2.1 Possible modes of transmission of the virus and associated symptoms of COVID-19.

Some racial and ethnic groups (Black, Hispanic, and Asian) were also more susceptible to SARS-CoV-2 virus infection (WHO 2020b; Cascella et al. 2021).

2.5 SYMPTOMS OF COVID-19

The symptoms of COVID-19 are fever, cough, sore throat, malaise, and myalgia, anosmia (inability to smell), ageusia (inability to taste), anorexia, nausea, diarrhoea, and shortness of breath. Dyspnea, tachypnea, and hypoxaemia are indicators of the severity of infection (Gandhi et al. 2020). He et al. (2021) studied the clinical symptoms in mild and severe COVID-19 patients. They categorised symptoms broadly into three categories: respiratory symptoms (fever, cough, dyspnea, expectorant, haemoptysis, chest tightness, and pharyngalgia); digestive symptoms (abdominal pain, diarrhoea, nausea/vomiting); and neurological symptoms (anorexia, fatigue, myalgia, and headache) shown in Figure 2.1. They found that symptoms like fever, anorexia, fatigue, cough, dyspnea, expectoration, haemoptysis, abdominal pain, and diarrhoea frequently occurred in severe cases (He et al. 2021).

2.6 MORTALITY STATISTICS

According to a WHO report, more than 6.4 million deaths to July 2022 were reported (WHO 2022a), however there is a discrepancy in the number of reported deaths and actual deaths. It is estimated that actual deaths are exceeded by 113% around the world, attributed to a lack of testing, a poor reporting system, and listing of other health reasons for COVID deaths (Buchholz 2021). The WHO also confirmed underreporting of COVID-19 deaths (WHO 2022b).

2.7 HEALTH CHALLENGES OF COVID-19

As we know, COVID-19 has a severe adverse effect on physical health. It hampers the normal physiological functioning of the body. Even after the recovery from immediate symptoms of the infection, the aftereffects and side effects are quite pronounced in many vulnerable individuals. Besides physical health, mental health and social health are equally affected by COVID-19, and the management approach should focus on all three dimensions of health in COVID-19 infection.

2.7.1 Mental Health in COVID-19

COVID-19 has not only disrupted the physiological health but also the mental health of millions of people across the globe. Both the COVID-19 victims who survived the dread of COVID-19 and non-infected people who lived under the threat of contracting COVID-19 have undergone the same level of mental distress equally. After the onset of COVID-19, the major emphasis was placed by governments, doctors, and scientists on the search for drugs and the development of vaccines to treat and prevent COVID-19 but the allied problem of mental distress was not given equal immediate attention, despite it being a major factor in determining overall health.

Research conducted throughout the world has shown that COVID-19 has devastated the mental wellbeing of every individual. Children, adolescents, adults, the elderly, caregivers, and health professionals have all experienced mental torment due to various reasons. According to a UNICEF report, the mental health of adolescents and young people has been affected by the COVID-19 crisis. In Latin America and the Caribbean, anxiety and depression were mainly reported among young people and adolescents due to financial stress. The respondents felt less motivated and less optimistic about future (UNICEF 2020). A study on the psychological impact of COVID-19 on adults living in the Middle East and North Africa region observed an increase in stress levels due to financial matters as well as domestic matters. The subjects felt horrified, helpless, and apprehensive due to the COVID-19 pandemic. High stress levels were experienced more by females, subjects aged 26–35 years, and those with lower education levels (Al Dhaheri et al. 2021; Choi et al. 2021).

A study of Australian adults showed that rates of anxiety and depression increased due to loneliness and financial distress (Dawel et al. 2020). Choi et al. (2021) also found a similar trend among Korean people who showed a decline in mental wellbeing and an increase in boredom. A study of university students showed a decline in physical activity and socialisation, an increase in sleeping hours and screen time, and an increase in depression. Disturbance in physical activity was associated with mental health (Giuntella et al. 2021). Similarly, children exhibited clingy behaviour, disturbed sleep, poor appetite, nightmares, inattentiveness, and separation problems (Singh et al. 2020). The sudden change in normal routine because of the closure of schools and daycare centres, confinement to homes, sudden introduction of online studies, and obligatory hygiene practices imposed by parents to maintain hygiene would have increased loneliness, fear, and anxiety levels among children. Distress was also prevalent among Indian people too. Young people, females, respondents working from workplaces rather than from home, who had no health insurance, a history of pre-existing medical conditions, and financial instability were more at risk (Anand et al. 2021). Older adults also experienced loneliness, depression, anxiety, and stress symptoms (Parlapani et al. 2021) due to social and emotional isolation, being unable to meet their friends, relatives, and attend social or religious gatherings, and being unable to cope with the new normal where digitalisation became mandatory, coupled with high vulnerability to infection. However, it was also observed that older people were better able to cope with stressors compared to younger people (Jiang 2020). Healthcare professionals and individuals infected with the virus who were under quarantine also experienced mental disturbance due to the stigma that they may spread the infection. They did not receive emotional and social support from the community during the initial phase of the pandemic. People under quarantine or healthcare professionals working with COVID-19 patients experienced feelings of sadness, anger, and frustration due to societal stigma. Continuous fear of contracting the disease, fear of losing loved ones, and the loss of those close to them influenced their psychological wellbeing. Quarantine and self-isolation have a negative impact on mental health (Javed et al. 2020). The psychological reactions to COVID-19 infection were manifested in specific and uncontrolled fears related to infections, pervasive anxiety due to uncertainty about the future, fear of infection leading to insomnia and depression, or post-traumatic stress. Frustration and boredom due to a change in normal routines, adapting to the new normal and social isolation had a detrimental effect on cognitive functioning and decision-making ability. It increased drug abuse, inability to socialise, helplessness, and suicidal behaviour (Serafini et al. 2020). Social isolation, reduced cognitive stimulation, and an increased risk of dementia were reported. Increased substance abuse to cope with stressors, suicidal ideation among youth and ethnic groups, anxiety, depression, trauma, and stress-related distress behaviours were common responses to pandemic (Czeisler et al. 2020).

2.7.2 Social Health in COVID-19

The unprecedented nature of COVID-19 has affected all dimensions of health. Besides physical and mental health, COVID-19 has also impinged on social health. Social health is defined as the ability to make healthy interpersonal relationships with others and coexist peacefully. The five dimensions of social health are social integration, social contribution, social coherence, social actualisation, and social acceptance (National Health Portal 2019).

It is recognised that social ties act as preventive medicine. Healthy supportive social relations improve physiological health as they have a positive effect on immune, cardiovascular, and endocrine functioning. Social support gives a sense of being loved, cared for, and valued which increases our ability to cope with stressors, leading to positive mental health. Healthy social relationships reduce morbidity and mortality risk (Umberson and Montez 2010).

During the initial phase of COVID-19, the only means to avoid the spread of the virus was through lockdown. Restricting the socialisation process and confining people to their homes were necessary steps, but somewhere in the process of controlling the COVID-19 infection, social health was jeopardised. The gravity of the impact of social confinement on health can be better understood as it is adopted as a technique of punishing criminals in jails. Self-isolation during COVID-19 affected social ties. Social distancing practised to prevent the spread of the virus had an adverse impact on social relationships, social interactions, and perceptions of empathy (Saladino et al. 2020). People in general, were not allowed to attend large gatherings, and many were even apprehensive of attending any social or religious functions, which affected their social and psychological wellbeing.

The only means of social connectedness were online platforms but they were far from the reach of those who were not technology-savvy or lacked the resources for online connectivity. The elderly, who are more dependent on their families for their functionality, were affected more by loneliness and social isolation (Hwang et al. 2020) probably because the elderly were less comfortable using online social platforms.

Besides the elderly, children and college-going students were also affected. Socialisation venues like schools, colleges, eating-out places, religious places, and family functions were closed during lockdown. Separation from peers had an impact on social, emotional, and cognitive development (Cameron and Tenenbaum 2021). Man, being a social animal, felt ostracised from society because of lockdown and social isolation. Difficulty in changing one's routine behaviour and conforming to the new standards of social behaviour (Krings et al. 2021) affected mental health and consequently physical health as well. Social connectedness is key to emerge from such a crisis. Research has proven that social participation has a positive effect on mental and physical health (Sepulveda-Loyola et al. 2020). Social support moderates the level of resilience and wellbeing. Social support, especially from family, acts as a stress-buffering component (Li et al. 2021). Social support from family and friends protects against depression and post-traumatic stress disorder. It has been seen that intimate friends and family are the strong ties and provide practical and emotional support but, on the contrary, sometimes increased social support may reduce self-efficacy (Liu et al. 2021). Engaging in social activity, having a large social network, and life space have a positive effect on cognitive functioning (Evans et al. 2019; Ingram et al. 2021). Physical, mental, and social health are interconnected, and none of the dimensions can be ignored for a healthy individual. Thus, the management strategy should focus on all three dimensions.

2.8 STRATEGIES FOR CONTROL AND TREATMENT

The core strategy for preventing and controlling COVID-19 is based on limiting the likelihood of transmission. COVID-appropriate behaviours, namely physical distancing, masking, maintaining hand hygiene, regular sanitisation of surroundings, and proper ventilation are the only means to prevent the infection which can be performed at an individual level. At the government level, intensifying testing, tracing, and treating, improving the healthcare system, restricting the movement of spreaders by making containment zones, confining gatherings, and imposing lockdowns, mass education, and accelerating the vaccination drive were used. These approaches were taken up by various countries worldwide to control the disaster created by COVID-19.

Though there is no confirmed pharmacological treatment for COVID-19, the drugs used for treatment are those which have proven effective in other viral diseases. The drug treatment of COVID-19 targets two levels. Firstly, treating the associated symptoms and sabotaging the viral replication cycle in the infected host. The probable mechanisms by which drugs have a therapeutic effect are blocking the entry of the virus into cells by inhibiting the glycosylation of host receptors, reducing disease progression of disease by targeting the S protein/ACE2 (angiotensin converting enzyme 2) interaction, inhibiting membrane fusion of the viral envelope, decreasing inflammatory response, treating cytokine release syndrome (Sanders et al. 2020). However, the drugs and adjunctive therapy like corticosteroids used for treating COVID-19 have been reported to cause side effects (Li et al. 2020; WHO 2020b). This calls for alternative and complementary therapies and dietary treatment to elevate immunity to prevent contraction of the infection and combat post-COVID symptoms. A few therapies to mitigate COVID-19 challenges are discussed in this section.

2.8.1 Diet/Food

Research shows contradictory changes in eating behaviour among various populations across the globe during COVID-19. Quarantine/lockdown had an influence on eating behaviour and lifestyle. Few studies have shown a positive effect on eating habits. People adopted a Mediterranean diet, consumed more fresh produce and homemade food, reduced alcohol intake and smoking (Di Renzo et al. 2020). On the other hand, many people responded negatively to lockdown and indulged in more consumption of junk food and unhealthy snacking, increased alcohol consumption and increased smoking (Husain and Ashkanani 2020; Giacalone et al. 2020; Sidor and Rzymski 2020). Bennett et al. (2021) reviewed studies and found that in the majority of studies an unhealthy pattern of eating was practised during lockdown. Snacking on comfort foods rich in energy and non-nutrients, especially at night which is not considered healthy, was seen. Respondents reported an increase in the number of meals per day and in portion size, thereby adding to the quantity of food intake leading to obesity, a common health concern seen during lockdown in most of the studies. The subjects skipped breakfast, consumed more ready-to-eat food which restricted diversity in the diet, influencing the nutritional status. However, in the same review, few studies informed favourable changes in dietary habits like an increase in the intake of fresh produce, home-cooked food, a decrease in the consumption of fast foods and alcohol, probably because of restrictions in eating out, time constraints, and less interaction with friends. A reduction in physical activity was a major health issue during lockdown. Thus, it can be inferred that for the majority of people, eating behaviour in terms of food quality and quantity was affected during the COVID-19 pandemic.

Past studies have opined that a balanced nutritious diet is important for general wellbeing and resisting any infection by strengthening the immune system. As we know, the SARS-CoV-2 virus seizes the immune system, and the stronger the immune system, the lower the chances of contracting the infection. Therefore, there is a need to adopt healthy eating habits, a balanced diet, and follow the principle of mindful eating.

Good nutrition is one of the important pillars of wellbeing for every individual, and it becomes more crucial during an infectious pandemic when immunity needs to be boosted to counteract the disease. Immunity helps to fight against pathogenic organisms. During an infection, the immune system activates, leading to an increase in metabolism. With the increase in metabolism, the energy requirement increases, which has to be compensated from energy-giving substrates like carbohydrates, protein, and fats through diet. Vitamins (A, D, E, C, folic acid, B6, B12) and minerals (zinc, selenium, copper, iron) are required for a strong immune system. They act as antioxidants, or a part of antioxidant enzymes, or regulators of gene expression in immune cells, or are needed for maturation, differentiation, and responsiveness of immune cells. Gut microbiota helps in regulating the immune system which becomes disrupted in an infection. The dysbiosis can be corrected by the intake of probiotics. The probiotics enhance innate immunity. In COVID-19 patients, dysbiosis was reported, which can be corrected with proper nutritional support. Inflammation due to cytokine syndrome can be controlled by n-3 fatty acids (Calder 2020; de Faria Coelho-Ravagnani 2021).

Besides a nutritious diet, food safety and good food practices are equally important to develop resilience towards disrupted outcomes during COVID-19. In designing a comprehensive approach to tackle dietary aspects during COVID-19, the importance of food safety and good food practices cannot be overlooked. Although there has not been much convincing evidence of the SARS-CoV-2 virus spreading through food, their ability to survive on paper, cardboard, plastic, metal, wood, and faecal-oral route transmission does not rule out the possibility of their presence on packaging material. Therefore, food safety at all stages of the food supply chain should be maintained. The contamination of food due to an infected person or cross-contamination may act as a fomite, and thus, the handling and consumption of contaminated food should be of prime concern during the pandemic. A food safety management system using Hazard Analysis Critical Control Point(HACCP) principles ensures food safety. Hygienic practices like the use of Personal Protective Equipment (PPE) kits, gloves, and masks by food handlers at every stage of the food chain should be followed. Disinfecting areas of contact, utensils, dishes, etc. should be practised. Vehicles used for the transportation of food should be sanitised and the required temperature should be maintained. It is known that the virus remains active in low temperatures and can survive for a long period of time if a low temperature is provided. Therefore, refrigerated and frozen foods should be handled with the utmost care before consumption. The need for packaging material with antiviral properties is very much required. Copper, silver, and zinc nanoparticle coating-based packaging material can be an answer to antiviral packaging material (Ceniti et al. 2021; Olaimat et al. 2020).

The third important component influencing nutritional health during the COVID era was food and nutrition security. Multiple studies testified to the food-insecure world during the COVID pandemic (Zhang et al. 2021; Lauren et al. 2021; Singh et al. 2021a; Parekh et al. 2021; Nguyen et al. 2021). The unprecedented nature of COVID-19 created fear among people and forced them to panic, buying and hoarding food, affecting the uniform availability of food for each segment. Besides, the lockdown jammed the supply from farm to consumer, affected the trade flow at the global and international levels, leading to poor accessibility and availability of food. The economic crisis due to loss of jobs reduced affordability. Food insecurity was experienced mainly by people in low socioeconomic strata, disadvantaged groups, the unemployed, and people with a lower level of education (Singh et al. 2021a; Parekh et al. 2021). Closing schools restricted children from benefiting from nutrition/feeding programmes, aggravating food and nutrition security among children (USGIC 2021). A decrease in food diversity as a result of a decrease in funds, low availability of food, and an increase in prices was seen (Nguyen et al. 2021), which led to skipping meals, eating less variety of food, and cutting portion size as a coping strategy by the affected group (Singh et al. 2021a). The COVID-19 pandemic has influenced food intake at all levels. Naja and Hamadeh (2020) gave a model that describes how the COVID-19 pandemic has influenced food and dietary intake at individual, community, national, and global levels. A check has to be made at each level to ensure food and nutrition security during a pandemic situation.

Therefore, in order to maintain physiological health, a constant supply of wholesome, nutritious, safe food for each individual is required. The three components, viz., immune-competent diet, food safety, and food security, as shown in Figure 2.2, should be addressed by governments, policymakers, and individuals for unprecedented pandemic preparedness.

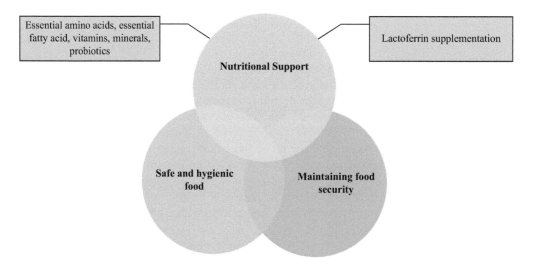

FIGURE 2.2   Components of maintaining an optimal nutritional state during a pandemic.

2.8.2 LACTOFERRIN

Lactoferrin, a non-toxic glycoprotein found in mammalian milk, has been seen to have antiviral, anti-inflammatory, and immunomodulatory effects. It has the potential to restrict the entry of the virus and its replication. Also, it modulates cytokines and decreases interleukin-6 and tumour necrosis factor-alpha, and thus the supplementation of lactoferrin can have a preventive and curative effect in COVID-19 (Chang et al. 2020). The iron-binding nature of lactoferrin makes less iron available for the replication of the SARS-CoV-2 virus. Besides its immune system stimulating effect, lactoferrin has a gastrointestinal protective effect too (Bell and Ormond 2022). Lactoferrin supplementation as an adjunctive therapy has been proven in few studies. In one of the studies conducted by Serrano et al. (2020), it was seen that liposomal lactoferrin supplementation along with vitamin C and zinc was effective in treating symptoms in COVID-19 subjects, and half the doses of supplementation had a preventive effect among their caregivers. Lactoferrin increases the rate of recovery, and a daily dose of 500–1000 mg of lactoferrin for treating COVID patients and half a dose of 250–500 mg daily for prevention has been suggested (Bell and Ormond 2022). Therefore, lactoferrin can be used as an adjunctive curative and preventive therapy in COVID-19 (Figure 2.2).

2.8.3 COMPLEMENTARY AND ALTERNATIVE THERAPY

Many herbs and plant products have been reported to possess antiviral, anti-inflammatory, and immune-boosting effects. The traditional knowledge of Ayurveda practised in India for ages has shown proven outcomes. The Ministry of AYUSH, Government of India, has given Ayurveda-based immunity-boosting measures for self-care during COVID-19 crisis, as shown in Figure 2.3. These Ayurvedic measures enhance immunity and provide relief from symptoms. Yoga practices improve respiratory and cardiac efficiency, enhance immunity, and reduce stress and anxiety, which are common mental health problems observed during the pandemic. The aqueous extracts made from various herbs are used as medication for prophylactic care as well as the treatment of mild COVID patients and post-COVID management (Ministry of AYUSH nd). The herbal treatments are reported to be safe and have no acute or severe side effects (Charan et al. 2021). Some of these spices and herbs are commonly used in the Indian kitchen.

Nugraha et al. (2020) concluded that certain herbs like echinacea, cinchona, curcumin, and xanthozzhizol had virucidal activity, immunomodulatory activity, antioxidative and anti-inflammatory effect but they have placed emphasis on preclinical and clinical trials for COVID-19. Herbal medicines like *Althaea officinalis*, *Commiphora molmol*, *Glycyrrhiza glabra*, *Hedera helix*, and *Sambucus nigra* are safe and have the potential of being used as adjuvant therapy for managing the symptoms of COVID-19 after well-defined pharmacological and clinical approval (Silveira et al. 2020). Natural plant products have certain bioactive compounds like flavonoids, alkaloids, peptides which possess antiviral (anti-SARS-CoV-2 activity) properties like ACE2, 3CLpro [3-chymotrypsin like protease], TMPRSS2 [transmembrane protease serine 2 inhibitor activity] (Antonio et al. 2020). The bioactive compound present in various spices and herbs may have promising outcome in the treatment and prevention of COVID-19. It has been seen that bioactive components present in curcumin, ginger, clove, black pepper, garlic, neem (*Azadirachta indica*), giloy (*Tinospora cordiflora*), basil, cinnamon have antiviral properties and enhance immunity (Singh et al. 2021b). The intake of these spices strengthens the defence system which is the major approach in beating COVID-19. Indian gooseberry (*amla*) and lemons were widely consumed during the COVID-19 pandemic in India, probably because they are rich sources of vitamin C which has antioxidant potential, a proven dietary supplement for fighting against COVID-19. Herbs like oregano, ginseng, sage, Indian ginseng (*ashwagandha*) have an immunomodulatory effect (Sharma 2020). In a review (Yashvardhini et al. 2021), it was reported that various molecular docking and *in silico* studies showed the role of various plants, herbs and spices like giloy (*Tinospora cordiflora*), ber (*Zizyphus jujube*), mahatita/

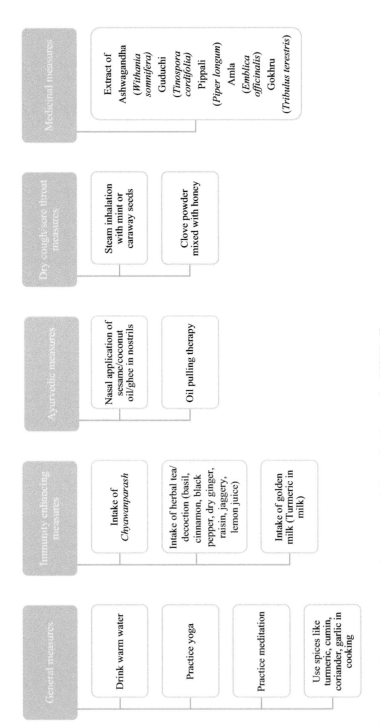

FIGURE 2.3   Ayurvedic measures as complementary and alternative therapy for COVID-19 management.

green chiretta (*Andrographis paniculate*), Indian cherry (*Cordia myxa*) and quince (*Cydonia oblonga*) in the prevention and prophylaxis of the coronavirus. Neem, tulsi, haldi, giloy, ginger, cloves, lemon, ashwagandha, and ginger have the potential to inhibit the SARS-CoV-2 infection. The possibilities of various Chinese herbs in the treatment of viral respiratory infection owing to the presence of anti-2019-nCoV compounds which may regulate inflammation reaction and hypoxia have been explored (Zhang et al. 2020). Epidemiological studies showing low COVID-19 death rates in the population consuming fermented vegetables open the door to the investigation of the potential role of fermented foods in controlling COVID-19. The probable mechanism underlying the contribution of fermented food like cabbage is the reduction in oxidative stress which occurs because of the binding of SARS-CoV-2 to ACE2. The fermented vegetables showed a potent anti-oxidant effect because of nuclear factor (erythoid-derived 2)-like 2 (Nrf2) (Bousquet et al. 2021a). Capsaicin present in red pepper, cinnamaldehyde in cinnamon, curcumin in turmeric, gingerol in ginger, piperine in black and long pepper and wasabi in Japanese horseradish were reported to show an Nrf2 activation effect (Bousquet et al. 2021b).

These studies show that phytochemicals present in some plants, herbs, spices, and condiments can be effectively used as supportive therapy in managing not only COVID-19 but other viral infections too. These herbs and spices can be used as potential ingredients for the development of antiviral functional foods in the future, as they have shown promising outcomes. Clinical studies are necessary to prescribe their safety levels. Thus, for preparedness against any unprecedented crisis in the future, more clinical research should be conducted, and governments need to lay down guidelines regarding the consumption of herbal medicines based on their efficacy and safety levels.

2.8.4 Lifestyle Modifications

The mental health repercussions stemming from social and biological reasons need to be attended to not only in unprecedented crises like COVID-19 but also generally. Mental health is an indispensable element of wellbeing and therefore needs special attention. Lifestyle plays a significant role in the maintenance of overall health. Talking specifically in reference to COVID-19, a disruption in lifestyle was observed, probably because of the sudden imposition of a lockdown. A decline in physical activity was the major lifestyle change due to the lockdown, which had an impact on mental, physiological, and social health. It was seen that physically inactive COVID-19 subjects had a greater risk of hospitalisation, ICU admission, and death (Sallis et al. 2021). A decrease in physical activity and an increase in sedentary behaviour like watching television and/or movies, playing computer and video games, and attending online classes was observed in children and adolescents (Dunton et al. 2020; Xiang et al. 2020). The largest decrease in physical activity was observed for teens, followed by older adults and young adults. The lower-educated and rural communities experienced a decrease in physical activity. Changes in earlier routines, lack of social support, and facilities were quoted as reasons for a decrease in physical activities (Schmidt and Pawlowski 2021). Physical activity has a positive effect on the immune system, cardio-metabolic and musculoskeletal health, improves sleep quality, cognitive abilities, mood, and vaccine response, and decreased systemic inflammation (Fuzeki et al. 2020; Damiot et al. 2020) and therefore an awareness needs to be created among the population to understand the importance of physical activity on physiological and mental health.

In the COVID-19 era, a change in alcohol consumption behaviour was also observed. Few studies showed a decrease in alcohol consumption because of a decline in social event participation. Hanging out with friends or partying were impeded because of the lockdown, and therefore the chances of drinking were also reduced. Also, a decline in availability and affordability would have restricted alcohol consumption. But many studies reported an increase in alcohol consumption. The respondents consumed alcohol as a means of coping with stress and anxiety, isolation, boredom, loss of daily routine, and loneliness imposed due to lockdown/social isolation, further aggravating physical and mental health (Calina et al. 2021). Similar reasons were found in other studies wherein a high percentage of respondents showed an increase in alcohol consumption because of stress, boredom, increased availability (Grossman et al. 2020), depression and anxiety (Tran et al. 2020). Binge drinking increased because of the long hours spent at home during lockdown (Weerakoon et al. 2021). Increased alcohol intake as a coping method for post-traumatic stress and depression abuse was also observed in hospital employees who worked in high-risk locations of contracting infection (Wu et al. 2008). An increase in alcohol consumption was found irrespective of gender (Pollard et al. 2020). As it is a known fact that alcohol consumption decreases immunity and increases physio-psychological imbalance, measures should be taken to abstain from alcohol consumption and follow healthy coping strategies for managing stress, anxiety, depression, loneliness, and isolation. Mindful meditation, exercise, and yoga should be adopted as a method for coping with psychological issues (Saeed et al. 2019; Lee 2007).

Smoking is another modifiable lifestyle factor which was observed to increase because of high mental stress during lockdown (Carreras et al. 2021). An increase in smoking because of loneliness and isolation was reported (Gendall et al. 2021). However, like alcohol consumption, smoking also showed a dual pattern of intake. Some subjects showed no change in smoking but many showed a rise in the frequency of smoking. Some respondents informed that they had quit tobacco because of its non-availability, increased price during lockdown, and association of tobacco with COVID-19 (Gupte et al. 2020). It has been found that active smoking and history of smoking were directly related to the severity of COVID-19 (Gulsen et al. 2020). Tobacco is harmful to the cardiovascular and respiratory system, and smokers are at a higher risk of COVID-19. Smoking is a risk factor for respiratory infection. It causes swelling and rupturing of air sacs in the lungs and reduces the lungs' capacity to exchange gases (WHO 2022c). Owing to the harmful effect of smoking and tobacco consumption, an awareness campaign should be carried out, and implementation of related policies should be ensured.

The uncertainty, fear, and stress of lockdown affected sleep patterns. Disturbed sleep patterns during lockdown were observed in many studies. Impaired sleep quality and sleep habits were seen (Franceschini et al. 2020). Sleeping late at night and waking late in the morning were observed (Agarwal et al. 2021). The majority of school-going children slept for 12 hours, but their sleeping time was around midnight (Ranjbar et al. 2021) indicating a disturbance in the sleep cycle. An increase in screen time and a decrease in physical activity during lockdown, especially among children and teens, affected the sleep pattern (Ranjbar et al. 2021). A decrease in physical activity and poor sleep quality is directly proportional to negative mood (Ingram et al. 2020). During the COVID-19 lockdown, a shift in sleep cycles led to more daytime napping, later bedtimes, and delayed onset of sleep, thereby increasing anxiety and depression (Gupta et al. 2020). Insomnia was observed due to loneliness, uncertainties, depression, and worry (Voitsidis et al. 2020). Irregular and disturbed sleeping patterns in COVID-19 lockdown restrictions should be taken up as a health emergency and the factors associated with disturbed sleep should be identified and controlled. It is important that awareness regarding sleep hygiene practices should be created. Small yet useful sleep hygiene practices like consistency in sleeping and waking times, a comfortable sleeping environment, avoiding heavy meals, caffeine, alcohol, turning off electronic devices before bedtime, and being physically active during the daytime should be practised (National Center for Chronic Disease Prevention and Health Promotion 2016).

It is the shared responsibility of citizens and governments to tackle unprecedented crises like COVID-19. The government should ensure that food, the most basic component, is within reach of each and every individual, especially during an emergency situation. The government should support the agriculture and food industries in times of crisis by waiving taxes, encouraging local produce, taking strict actions to maintain hygienic conditions at each level of the food supply chain, discouraging hoarding and panic buying, identifying malnourished areas, and ensuring the availability of essential commodities to the needy population. A social protection system should be developed and strengthened to facilitate the surveillance and intervene in social issues like domestic violence and child abuse which have been found to be on the rise (Pereda and Diaz-Faes 2020; Evans et al. 2020; Sharma and Borah 2022). Government, non-government organisations, and health workers should create awareness among the population. Authentic and genuine information should be disseminated to avoid panic among citizens. Priority should be given to strengthening the healthcare system, and every citizen should be covered under a health insurance scheme. Counselling centres at schools, colleges, workplaces and other institutions should be established both offline and online to provide psychological guidance and social help to those in need. Courses related to coping skills and crisis management should become an integral part of the curriculum to prepare children from a young age for any crisis in the future. Telemedicine, telepsychology, online consultations, and online social activities can be provided by responsible agencies. Citizens should support the government and play their part (Figure 2.4) and also follow practical tips given by various health organisations (Table 2.1) to come out of crisis collectively.

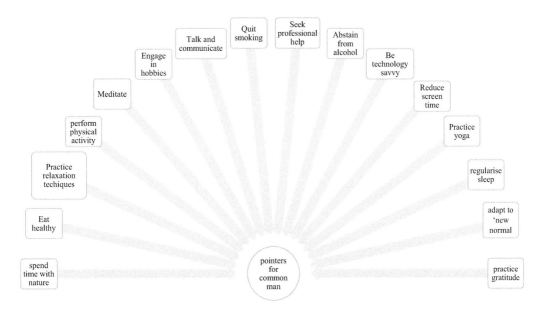

FIGURE 2.4   Common pointers to follow for management of COVID-19.

TABLE 2.1
Practical Tips for Maintaining Health during a Crisis

Dos	Don'ts
Plan and purchase what is needed.	Panic buy.
Prioritise fresh produce.	Hoard food.
Prefer home cooked healthy meals.	Eat out/takeaways under unhygienic conditions.
Understand your healthy portion size.	Overeat.
Follow safe handling practices.	Cross contaminate.
Limit salt/sugar/saturated fat intake.	Consume excessive salt/sugar/saturated fat.
Keep hydrated.	Consume tea/coffee/caffeinated drinks/alcohol.
Consume fibre.	Consume refined grains.
Cook food at the right temperature.	Undercook or eat raw food.

2.9 CONCLUSION

COVID-19 is one of the worst humanitarian crises experienced in the 21st century. It is a contagious disease caused by severe acute respiratory syndrome coronavirus-2. With no confirmed pharmacological treatment yet available, social distancing, masking, sanitising, and vaccination are a few preventive measures to control the spread of the deadly disease. The unknown virus has caused mayhem globally since December 2019, after its identification for the first time. Millions of people lost their lives and livelihoods to the unprecedented crisis. It created havoc in the lives of every individual jeopardising health and wellbeing. Not only physiological health but mental and social health were also negatively influenced, which can never be indemnified. The lockdown imposed as the necessary step to restrict the virus from spreading actually had an adverse effect on physical, mental, and social health. The uncertainties associated with the pandemic like COVID-19 need to be tackled by strengthening our sociopolitical machinery. Preparedness on the part of governments and citizens is essential to combat any such disaster in the future. Health scientists and nutritionists can play a vital role in developing functional foods based on established herbs and spices that would be safe to improve immunity and fight the associated symptoms of not only COVID-19 infection but any such immunity comprising infection in the future. An integrated approach comprising awareness and interventions at each level is required. Policies to ensure a secure food system, an efficient healthcare system, an effective social support system, psychological resilience system, and financial stability are needed to address the health damage caused by the COVID pandemic.

REFERENCES

Agarwal, V., D. Goel, V.K. Paliwal, L. Gupta, A. Ghodke, and V. Agarwal. 2021. Disruptions in sleep due to prolonged isolation during covid-19 lockdown: A survey based study. *Journal of Sleep Disorders and Management* 7: 032. https://doi.org/10.23937/2572-4053.1510032

Al Dhaheri, A.S., M.F. Bataineh, M.N. Mohamad, et al. 2021. Impact of COVID-19 on mental health and quality of life: Is there any effect? A cross-sectional study of the MENA region. *Plos One* 16, no. 3: e0249107. https://doi.org/10.1371/journal.pone.0249107

Anand, V., L. Verma, A. Aggarwal, P. Nanjundappa, and H. Rai. 2021. COVID-19 and psychological distress: Lessons for India. *Plos One* 16, no. 8: e0255683. https://doi.org/10.1371/journal.pone.0255683

Antonio, A., L.S.M. Wiedemsnn, and V.F. Veiga-Junior. 2020. Natural products' role against COVID-19. *RSC Advances* 10: 23379–23393.

Bell, S.J., and L.R. Ormond. 2022. Lactoferrin for treating and preventing covid-19: A review. *Acta Scientific Nutritional Health* 6: 61–67.

Bennett, G., E. Young, I. Butler, and S. Coe. 2021. The impact of lockdown during the COVID-19 outbreak on dietary habits in various population groups: A scoping review. *Frontiers in Nutrition* 8: 626432. https://doi.org/10.3389/fnut.2021.626432

Bousquet, J., J.M. Anto, W. Czarlewski, et al. 2021a. Cabbage and fermented vegetables: From death rate heterogeneity in countries to candidates for mitigation strategies of severe COVID-19. *Allergy* 76, no. 3: 735–750. https://doi.org/10.1111/all.14549

Bousquet, J., W. Czarlewski, T. Zuberbier, et al. 2021b. Spies to control COVID-19 symptoms: Yes, but not only.... *International Archives of Allergy and Immunology* 182: 489–495. https://doi.org/10.1159/000513538

Buchholz, K. 2021. Chart: The global problem of underreporting COVID-19 deaths. *The Wire.* 13 May 2021. https://thewire.in/health/chart-the-global-problem-of-underreporting-covid-19-deaths

Calder, P.C. 2020. Nutrition, immunity and COVID-19. *BMJ Nutrition, Prevention & Health* 3: e000085. https://doi.org/10.1136/bmjnph-2020-000085

Calina, D., T. Hartung, I. Mardare, et al. 2021. COVID-19 pandemic and alcohol consumption: Impacts and interconnections. *Toxicology Reports* 8: 529–535. https://doi.org/10.1016/j.toxrep.2021.03.005

Cameron, L., and H. Tenenbaum. 2021. Lessons from developmental science to mitigate the effects of the COVID-19 restrictions on social development. *Group Processes and Intergroup Relations* 24: 231–236.

Carreras, G., A. Lugo, C. Stival, et al. 2021. Impact of COVID-19 lockdown on smoking consumption in a large representative sample of Italian adults. *Tobacco Control.* https://doi.org/10.1136/tobaccocontrol-2020-056440

Cascella, M., M. Rajnik, A. Aleem, S.C. Dulebohn, and R. Di Napoli. 2021. *Features, Evaluation, and Treatment of Coronavirus (COVID-19)*. Treasure Island (FL): StatPearls Publishing. www.ncbi.nlm.nih.gov/books/NBK554776/

Ceniti, C., B. Tilocca, D. Britti, A. Santoro, and N. Costanzo. 2021. Food safety concerns in "COVID-19 Era". *Microbiological Research* 12: 53–68. https://doi.org/10.3390/microbiolres12010006

Chang, R., T.B. Ng, and W.Z. Sun. 2020. Lactoferrin as potential preventative and adjunct treatment for COVID-19. *International Journal of Antimicrobial Agents* 56, no. 3: 106118. https://doi.org/10.1016/j.ijantimicag.2020.106118

Charan, J., P. Bhardwaj, S. Dutta, et al. 2021. Use of complementary and alternative medicine (cam) and home remedies by COVID-19 patients: A telephonic survey. *Indian Journal of Clinical Biochemistry* 36, no. 1: 108–111.

Choi, I., J.H. Kim, N. Kim, et al. 2021. How COVID-19 affected mental well-being: An 11- week trajectories of daily well-being of Koreans amidst COVID-19 by age, gender and region. *Plos One* 16, no. 4: e0250252. https://doi.org/10.1371/journal.pone.0250252

Czeisler, M.E., R.I. Lane, E. Petrosky, et al. 2020. Mental health, substance use, and suicidal ideation during the COVID-19 pandemic – United States. *Morbidity and Mortality Weekly Report* 69: 1049–1057. https://doi.org/10.15585/mmwr.mm6932a1

Damiot, A., A.J. Pinto, J.E. Turner, and B. Gualano. 2020. Immunological implications of physical inactivity among older adults during COVID-19 pandemic. *Gerontology* 66: 431–438.

Dawel, A., Y. Shou, M. Smithson, et al. 2020. The effect of COVID-19 on mental health and wellbeing in a representative sample of Australian adults. *Front Psychiatry* 11: 579985. https://doi.org/10.3389/fpsyt.2020.579985

de Faria Coelho-Ravagnani, C., F.C. Corgosinho, F.F.Z. Sanches, C.M.M. Prado, A. Laviano, and J.F. Mota. 2021. Dietary recommendations during the COVID-19 pandemic. *Nutrition Reviews* 79, no. 4: 382–393.

Di Renzo, L., P. Gualtieri, F. Pivari, et al. 2020. Eating habits and lifestyle changes during COVID-19 lockdown: An Italian survey. *Journal of Translational Medicine* 18: 229. https://doi.org/10.1186/s12967-020-02399-5

Drexler, M. 2010. *What You Need to Know about Infectious Disease*. Washington (DC): National Academies Press. www.ncbi.nlm.nih.gov/books/NBK209706

Dunton, G.F., B. Do, and S.D. Wang. 2020. Early effects of the COVID-19 pandemic on physical activity and sedentary behavior in children living in the U.S. *BMC Public Health* 20: 1351. https://doi.org/10.1186/s12889-020-09429-3

Evans, I., A. Martyr, R. Collins, C. Brayne, and L. Clare. 2019. Social isolation and cognitive function in later life: A systematic review and meta-analysis. *Journal of Alzheimer's Disease* 70, no. s1: S119–S144. https://doi.org/10.3233/JAD-180501

Evans, M.L., M. Lindauer, and M.E. Farrell. 2020. A pandemic within pandemic-intimate partner violence during Covid-19. *New England Journal of Medicine* 383: 2302–2304.

Franceschini, C., A. Musetti, C. Zenesini, et al. 2020. Poor sleep quality and its consequences on mental health during the COVID-19 lockdown in Italy. *Frontiers in Psychology* 11: 574475. https://doi.org/10.3389/fpsyg.2020.574475

Fuzeki, E., D.A. Groneberg, and W. Banzer. 2020. Physical activity during COVID-19 induced lockdown: Recommendations. *Journal of Occupational Medicine and Toxicology* 15: 25. https://doi.org/10.1186/s12995-020-00278-9

Gandhi, R.T., J.B. Lynch, and C.D. Rio. 2020. Mild or moderate Covid-19. *New England Journal of Medicine* 383: 1757–1766.

Gendall, P., J. Hoek, J. Stanley, M. Jenkins, and S. Every-Palmer. 2021. Changes in tobacco use during the 2020 COVID-19 lockdown in New Zealand. *Nicotine & Tobacco Research* 23, no. 5: 866–871. https://doi.org/10.1093/ntr/ntaa257

Giacalone, D., M.B. Frost, and C. Rodríguez-Perez. 2020. Reported changes in dietary habits during the COVID-19 lockdown in the Danish population: The Danish COVIDiet study. *Frontiers in Nutrition* 7: 592112. https://doi.org/10.3389/fnut.2020.592112

Giuntella, O., K. Hyde, S. Saccardo, and S. Sadoff. 2021. Lifestyle and mental health disruptions during COVID-19. *Proceedings of the National Academy of Sciences* 118, no. 9: e2016632118. https://doi.org/10.1073/pnas.2016632118

Grossman, E.R., S.E. Benjamin-Neelon, and S. Sonnenschein. 2020. Alcohol consumption during the COVID-19 pandemic: A cross-sectional survey of US adults. *International Journal of Environmental Research and Public Health* 17, no. 24: 9189. https://doi.org/10.3390/ijerph17249189

Gulsen, A., B.A. Yigitbas, B. Uslu, D. Dromann, and O. Kilinc. 2020. The effect of smoking on COVID-19 symptom severity: Systematic review and meta-analysis. *Pulmonary Medicine*: 7590207. https://doi.org /10.1155/2020/7590207

Gupta, R., S. Grover, A. Basu, et al. 2020. Changes in sleep pattern and sleep quality during COVID-19 lockdown. *Indian Journal of Psychiatry* 62, no. 4: 370–378. https://doi.org/10.4103/psychiatry

Gupte, H.A., G. Mandal, and D. Jagiasi. 2020. How has the COVID-19 pandemic affected tobacco users in India: Lessons from an ongoing tobacco cessation program. *Tobacco Prevention & Cessation* 6: 53. https://doi.org/10.18332/tpc/127122

He, X., X. Cheng, X. Feng, H. Wan, S. Chen, and M. Xiong. 2021. Clinical symptom differences between mild and severe COVID-19 patients in China: A meta-analysis. *Frontiers in Public Health* 8: 561264. https:// doi.org/10.3389/fpubh.2020.561264

Huang, Y., C. Yang, X.-F. Xu, W. Xu, and S.W. Liu. 2020. Structural and functional properties of SARS-CoV-2 spike protein: Potential antivirus drug development for COVID-19. *Acta Pharmacologica Sinica* 41: 1141–1149. https://doi.org/10.1038/s41401-020-0485-4

Husain, W., and F. Ashkanani. 2020. Does COVID-19 change dietary habits and lifestyle behaviours in Kuwait: A community-based cross-sectional study. *Environmental Health and Preventive Medicine* 25: 61. https://doi.org/10.1186/s12199-020-00901-5

Hwang, T.J., K. Rabheru, C. Peisah, W. Reichman, and M. Ikeda. 2020. Loneliness and social isolation during the COVID-19 pandemic. *International Psychogeriatrics* 32, no. 10: 1217–1220. https://doi.org/10.1017 /S1041610220000988

Ingram, J., C.J. Hand, and G. Maciejewski. 2021. Social isolation during COVID-19 lockdown impairs cognitive function. *Applied Cognitive Psychology* 35, no. 4: 935–947.

Ingram, J., G. Maciejewski, and C.J. Hand. 2020. Changes in diet, sleep, and physical activity are associated with differences in negative mood during COVID-19 lockdown. *Frontiers in Psychology* 11: 588604. https://doi.org/10.3389/fpsyg.2020.588604

Javed, B., A. Sarwer, E.B. Soto, and Z-R. Mashwani. 2020. The coronavirus (COVID-19) pandemic's impact on mental health. *International Journal of Health Planning and Management* 35, no. 5: 993–996. https://doi.org/10.1002/hpm.3008

Jiang, D. 2020. Perceived stress and daily well-being during the COVID-19 outbreak: The moderating role of age. *Frontiers in Psychology* 11: 571873. https://doi.org/10.3389/fpsyg.2020.571873

Krings, V.C., B. Steeden, D. Abrams, and M.A. Hogg. 2021. Social attitudes and behaviours in the COVID-19 pandemic: Evidence and prospects from research on group processes and intergroup relations. *Group Processes & Intergroup Relations* 24, no. 2: 195–2021.

Lauren, B.N., E.R. Silver, A.S. Faye, et al. 2021. Predictors of households at risk for food insecurity in the United States during the COVID-19 pandemic. *Public Health Nutrition* 24, no. 12: 3929–3936. Epub 2021. https://doi.org/10.1017/S1368980021000355

Lee, S.H., S.C. Ahn, Y.J. Lee, T.K. Choi, K.H. Yook, and S.Y. Suh. 2007. Effectiveness of a meditation-based stress management program as an adjunct to pharmaco therapy in patients with anxiety disorder. *Journal of Psychosomatic Research* 62, no. 2: 189–195.

Li, F., S. Luo, W. Mu, et al. 2021. Effects of sources of social support and resilience on the mental health of different age groups during the COVID-19 pandemic. *BMC Psychiatry* 21: 16. https://doi.org/10.1186/ s12888-020-03012-1

Li, L., R. Li, Z. Wu, et al. 2020. Therapeutic strategies for critically ill patients with COVID-19. *Annals of Intensive Care* 10: 45. https://doi.org/10.1186/s13613-020-00661-z

Liu, C., N. Huang, M. Fu, H. Zhang, X.L. Feng, and J. Guo. 2021. Relationship between risk perception, social support, and mental health among general Chinese population during the COVID-19 pandemic. *Risk Management and Healthcare Policy* 14: 1843–1853.

Michaud, C.M. 2009. Global burden of infectious diseases. *Encyclopedia of Microbiology*: 444–454. https:// doi.org/10.1016/B978-012373944-5.00185-1

Ministry of AYUSH. n.d. National Clinical Management protocol based on ayurveda and yoga for management of COVID-19. Ministry of AYUSH, New Delhi. https://www.ayush.gov.in/docs/ayush-Protocol -covid-19.pdf (accessed 8 August 2022).

Murray, C.J., and A.D. Lopez. 1997. Mortality by cause for eight regions of the world: Global Burden of Disease Study. *Lancet* 349, no. 9061: 1269–1276. https://doi.org/10.1016/S0140-6736(96)07493-4

Naja, F., and R. Hamadeh. 2020. Nutrition amid the COVID-19 pandemic: A multi-level framework for action. *European Journal of Clinical Nutrition* 74: 1117–1121. https://doi.org/10.1038/s41430-020-0634-3

National Center for Chronic Disease Prevention and Health Promotion. 2016. Tips for better sleep. www.cdc .gov/sleep/about_sleep/sleep_hygiene.html

National Health Portal. 2019. Social health. https://www.nhp.gov.in/social-health_pg (accessed 8 August 2022).

Nguyen, P.H., S. Kachwaha, A. Pant, et al. 2021. Impact of COVID-19 on household food insecurity and interlinkages with child feeding practices and coping strategies in Uttar Pradesh, India: A longitudinal community-based study. *BMJ Open* 11: e048738. https://doi.org/10.1136/bmjopen-2021-048738

Nugraha, R.V., H. Ridwansyah, M. Ghozali, A.F. Khairani, and N. Atik. 2020. Traditional herbal medicine candidates as complementary treatments for COVID-19: A review of their mechanisms, pros and cons. *Evidence-Based Complementary and Alternative Medicine* 2020: 2560645. https://doi.org/10.1155/2020/2560645

Olaimat, A.N., H.M. Shahbaz, N. Fatima, S. Munir, and R.A. Holley. 2020. Food safety during and after the era of COVID-19 pandemic. *Front Microbiology* 11: 1854. https://doi.org/10.3389/fmicb.2020.01854

Parekh, N., S.H. Ali, J. O'Connor, et al. 2021. Food insecurity among households with children during the COVID-19 pandemic: Results from a study among social media users across the United States. *Nutrition Journal* 20: 73. https://doi.org/10.1186/s12937-021-00732-2

Parlapani, E., V. Holeva, V.A. Nikopoulou, S. Kaprinis, I. Nouskas, and I. Diakogiannis. 2021. A review on the COVID-19-related psychological impact on older adults: Vulnerable or not? *Aging Clinical and Experimental Research* 33: 1729–1743. https://doi.org/10.1007/s40520-021-01873-4

Pereda, N., and D.A. Díaz-Faes. 2020. Family violence against children in the wake of COVID-19 pandemic: A review of current perspectives and risk factors. *Child and Adolescent Psychiatry and Mental Health* 14: 40. https://doi.org/10.1186/s13034-020-00347-1

Pollard, M.S., J.S. Tucker, and H.D. Green. 2020. Changes in adult alcohol use and consequences during the COVID-19 pandemic in US. *JAMA Network Open* 3, no. 9: e2022942. https://doi.org/10.1001/jamanetworkopen.2020.22942

Ranjbar, K., H. Hosseinpour, R. Shahriarirad, et al. 2021. Students' attitude and sleep pattern during school closure following COVID-19 pandemic quarantine: A web-based survey in south of Iran. *Environmental Health and Preventive Medicine* 26, 33. https://doi.org/10.1186/s12199-021-00950-4

Saeed, A.S., K. Cunningham, and R.M. Bloch. 2019. Depressive and anxiety disorders: Benefits of exercise, yoga and meditation. *American Family Physician* 99, no. 10: 620–627.

Saladino, V., D. Algeri, and V. Auriemma. 2020. The psychological and social impact of covid-19: New perspectives of well-being. *Frontiers in Psychology* 11: 577684. https://doi.org/10.3389/fpsyg.2020.577684

Sallis, R., D.R. Young, S.Y. Tartof, et al. 2021. Physical inactivity is associated with a higher risk for severe COVID-19 outcomes: A study in 48 440 adult patients. *British Journal of Sports Medicine* 55: 1099–1105.

Sanders, J.M., M.L. Monogue, T.Z. Jodlowski, and J.B. Cutrell. 2020. Pharmacologic treatments for coronavirus disease 2019 (COVID-19): A review. *JAMA* 323, no. 18: 1824–1836. https://doi.org/10.1001/jama.2020.6019

Schmidt, T., and C.S. Pawlowski. 2021. Physical activity in crisis: The impact of COVID-19 on Danes' physical activity behavior. *Frontiers in Sports and Active Living* 2: 610255. https://doi.org/10.3389/fspor.2020.610255

Sepulveda-Loyola, W., I. Rodriguez-Sanchez, P. Perez-Rodríguez, et al. 2020. Impact of social isolation due to covid-19 on health in older people: Mental and physical effects and recommendations. *Journal of Nutrition, Health and Aging* 24, no. 9: 938–947. https://doi.org/10.1007/s12603-020-1469-2

Serafini, G., B. Parmigiani, A. Amerio, A. Aguglia, L. Sher, and M. Amore. 2020. The psychological impact of COVID-19 on the mental health in the general population. *QJM: An International Journal of Medicine* 113, no. 80: 531–537. https://doi.org/10.1093/qjmed/hcaa201

Serrano, G., I. Kochergina, A. Albors, et al. 2020. Liposomal lactoferrin as potential preventative and cure for COVID-19. *International Journal of Health Sciences and Research* 8, no. 1: 8–15.

Sharma, A., and S.B. Borah. 2022. Covid-19 and domestic violence: An indirect path to social and economic crisis. *Journal of Family Violence* 37: 759–765. https://doi.org/10.1007/s10896-020-00188-8

Sharma, L. 2020. Immunomodulatory effect and supportive role of traditional herbs, spices and nutrients in management of COVID-19. Preprint. https://doi.org/10.20944/preprints202009.0026.v1

Sidor, A., and P. Rzymski. 2020. Dietary choices and habits during COVID-19 lockdown: Experience from Poland. *Nutrients* 12, no. 6: 1657. https://doi.org/10.3390/nu12061657

Silveira, D., J.M. Prieto-Garcia, F. Boylan, et al. 2020. COVID-19: Is there evidence for the use of herbal medicines as adjuvant symptomatic therapy? *Frontiers in Pharmacology* 11: 581840. https://doi.org/10.3389/fphar.2020.581840

Singh, D.R., D.R. Sunuwar, S.K. Shah, L.K. Sah, K. Karki, and R.K. Sah. 2021a. Food insecurity during COVID-19 pandemic: A genuine concern for people from disadvantaged community and low-income families in Province 2 of Nepal. *Plos One* 16, no. 7: e0254954. https://doi.org/10.1371/journal.pone.0254954

Singh, N.A., P. Kumar, Jyoti, and N. Kumar. 2021b. Spices and herbs: Potential antiviral preventives and immunity boosters during COVID-19. *Phytotherapy Research* 2021: 1–13. https://doi.org/10.1002/ptr .7019

Singh, S., D. Roy, K. Sinha, S. Parveen, G. Sharma, and G. Joshi. 2020. Impact of COVID-19 and lockdown on mental health of children and adolescents: A narrative review with recommendations. *Psychiatry Research* 293: 113429. https://doi.org/10.1016/j.psychres.2020.113429

Smith, K.F., M. Goldberg, S. Rosenthal, et al. 2014. Global rise in human infectious disease outbreaks. *Journal of The Royal Society Interface* 11, no. 101: 20140950. https://doi.org/10.1098/rsif.2014.0950

Tran, T.D., K. Hammarberg, M. Kirkman, H.T.M. Nguyen, and J. Fisher. 2020. Alcohol use and mental health status during the first months of COVID-19 pandemic in Australia. *Journal of Affective Disorders* 277: 810–813. https://doi.org/10.1016/j.jad.2020.09.012

Umberson, D., and J.K. Montez. 2010. Social relationships and health: A flashpoint for health policy. *Journal of Health and Social Behavior* 51: S54–S66. https://doi.org/10.1177/0022146510383501

UNICEF. 2020. The impact of COVID-19 on the mental health of adolescents and youth. https://www.unicef .org/lac/en/impact-covid-19-mental-health-adolescents-and-youth

USGIC. 2021. COVID-19 brief: Impact on food security 2021. www.usglc.org/coronavirus/global-hunger (accessed 31 July 2022).

van Seventer, J.M., and N.S. Hochberg. 2017. Principles of infectious diseases: Transmission, diagnosis, prevention, and control. *International Encyclopedia of Public Health* 6: 22–39. https://doi.org/10.1016/ B978-0-12-803678-5.00516-6

Voitsidis, P., I. Gliatas, V. Bairachtari, et al. 2020. Insomnia during the COVID-19 pandemic in a Greek population. *Psychiatry Research* 289: 113076. https://doi.org/10.1016/j.psychres.2020.113076

Weerakoon, S.M., K.K. Jetelina, and G. Knell. 2021. Longer time spent at home during COVID-19 pandemic is associated with binge drinking among US adults. *The American Journal of Drug and Alcohol Abuse* 47, no. 1: 98–106. https://doi.org/10.1080/00952990.2020.1832508

WHO. 2020a. Transmission of SARS-CoV-2: Implication for infection prevention precautions. www.who.int/ news-room/commentaries/detail/transmission-of-sars-cov-2-implications-for-infection-prevention-precautions (accessed 30 May 2022).

WHO. 2020b. Clinical management of COVID-19. *Interim Guidance*. 27 May 2020. WHO. https://apps.who .int/iris/handle/10665/332196 (accessed 20 April 2022).

WHO. 2022c. Tobacco and waterpipe use increases the risk of COVID-19. www.emro.who.int/tfi/know-the -truth/tobacco-and-waterpipe-users-are-at-increased-risk-of-covid-19-infection.html

WHO. 2022a. WHO coronavirus (COVID-19) dashboard. https://covid19.who.int (accessed 8 August 2022).

WHO. 2022b. The true death toll of COVID-19: Estimating global excess mortality. https://www.who.int/data /stories/the-true-death-toll-of-covid-19-estimating-global-excess-mortality (accessed 8 August 2022).

Wu, P., X. Liu, Y. Fang, et al. 2008. Alcohol abuse/dependence symptoms among hospital employees exposed to a SARS outbreak. *Alcohol Alcohol* 43, no. 6: 706–712. https://doi.org/10.1093/alcalc/agn073

Xiang, M., Z. Zhang, and K. Kuwahara. 2020. Impact of COVID-19 pandemic on children and adolescents' lifestyle behavior larger than expected. *Progress in Cardiovascular Diseases* 63, no. 4: 531–532. https:// doi.org/10.1016/j.pcad.2020.04.013

Yashvardhini, N., S. Samiksha, and D. Jha. 2021. Pharmacological intervention of various Indian medicinal plants in combating COVID-19 infection. *Biomedical Research and Therapy* 8, no. 7: 4461–4475. https://doi.org/10.15419/bmrat.v8i7.685

Zhang, D.H., K.L. Wu, X. Zhang, S.Q. Deng, and B. Peng. 2020. In silico screening of Chinese herbal medicines with the potential to directly inhibit 2019 novel coronavirus. *Journal of Integrative Medicine* 18, no. 2: 152–158.

Zhang, Y., K. Yang, S. Hou, T. Zhong, and J. Crush. 2021. Factors determining household-level food insecurity during COVID-19 epidemic: A case of Wuhan, China. *Food & Nutrition Research* 65. https://doi.org /10.29219/fnr.v65.5501

3 Undernutrition and Hidden Hunger

3.1 INTRODUCTION

Despite the advances being made in every sphere of development, undernutrition has remained a major health challenge for a large segment of the population throughout the world. Simply defined, undernutrition is a state when an insufficient amount of calories and nutrients is being consumed by an individual. The inadequate intake of food and/or intake of poor-quality food lead to undernutrition. The repercussions of undernutrition in the form of its linkages with other morbidities and mortality pose a gigantic health threat to mankind at any stage of the lifespan.

Undernutrition is classified into four groups: underweight, stunting, wasting, and hidden hunger. It is the state when an individual has low weight for his/her age, stunting refers to low height for age, wasting is low weight for height, and hidden hunger is a form of undernutrition when the intake of micronutrients is not sufficient.

3.2 MAGNITUDE OF UNDERNUTRITION AND HIDDEN HUNGER

Undernutrition and hidden hunger have pervaded the whole world, hitting Asia and Africa the most. Regardless of the immense efforts being made under the Sustainable Development Goals to end hunger, achieve food security, and improve nutrition, little progress is seen at present, jeopardised by sudden crises like COVID-19, wars, climate change, etc. According to a United Nations report, it was estimated that in 2021 globally 828 million people were suffering from hunger and nearly one in three lacked regular access to adequate food. Moderate or severe food insecurity was experienced by 2.3 billion people, affecting women (31.9%) more than men (27.6%). The data are sufficient to reflect that the zero hunger goal is at risk, as it is estimated that more than 600 million people will face hunger in 2030 (United Nations 2022a; WHO 2022). According to World Food Programme (WFP) estimates, more than 345 million people face high levels of food insecurity in 2023, which is more than double the number in 2020 (wfp.org/global-hunger-crisis).

Undernutrition persists worldwide, affecting the wellbeing of vulnerable populations, mostly children under 5. According to UNICEF/WHO/World Bank Group Joint Child Malnutrition Estimates (2023), undernutrition was the cause of nearly half of all deaths among children under 5 in 2022. It was reported that in 2022, 148.1 million (22.3%) children under 5 were stunted and 45 million (6.8%) were wasted, of which 52% of stunted and 70% of wasted children live in Asia. Additionally, 43% of stunted and 27% of wasted children live in Africa. It is projected that if the rate of current progress in tackling undernutrition among children continues, then 128.5 million (19.5%) children will still be stunted in 2030.

3.3 CAUSES OF UNDERNUTRITION AND HIDDEN HUNGER

3.3.1 INDIVIDUAL LEVEL

Undernutrition and hidden hunger can ensue at different stages of life with the most severe consequences seen during the period from conception until 2 years of age. Unborn and newborn babies are at higher risk if the mothers are undernourished, anaemic, or underfed or suffering from infections or incomplete breastfeeding practices. However, during the weaning period, there are chances

DOI: 10.1201/9781003354024-3

of low dietary intake or imbalance in the intake of nutrients (Keeley et al. 2019). Another key factor is infections and infestations contracted by the child that may hinder the proper absorption of nutrients resulting in undernutrition.

Adolescence is the second window of opportunity, after the first 1000 days, to invest in nutritional interventions to avoid further malnutrition. In growing children and youths, the nutritional requirements increase. However, due to peer pressure and food fads and consumption of a generally nutrient-poor diet, many of them do not attain their nutritional needs, which might result in deficiencies (Burgess 2008).

During adulthood, several factors influence the nutritional status of an individual. Lack of physical activity, changes in appetite, work atmosphere, health status, and the extent of hospitalisations has a direct impact on the nutritional adequacy of adults.

3.3.2 Household Level

The characteristics of a family play a crucial role in determining the nutritional status of its members. The quality of the diet is dependent on the affordability of the household. Family income and family size are direct determinants of the nutritional sufficiency of a household. The larger the family size and lower the economic status of the family, the higher are the chances of the prevalence of undernutrition in the family.

Social and cultural practices adopted by a family greatly influence the household nutritional status. Traditional beliefs and cultural practices of a community constrain them from consumption of certain groups of foods. Similarly, many social groups have taboos and myths and misinterpretations which may cause serious deficiencies. The knowledge level of the caregivers or family plays a crucial role as well. The ignorance of household members can lead to insufficient food choices.

Child-rearing practices are also highly influenced by the beliefs, attitudes, and habits of the households. The feeding, nutrition, hygiene, sleep, clothing, discipline, habit training, etc. have direct consequences on the child's nutritional status.

3.3.3 Maternal Factors

The pregnancies of a woman deeply impact her nutritional status. Ignorance or lack of accessibility to information and resources, or family conditions, often lead to pregnancies that are too early, too late, too many, or too closely spaced. The direct impact of these pregnancies is seen not only on the mother but on the child as well. The most common manifestations of these circumstances are children with low birth weight (infants born with a birth weight less than 2.5 kg) and anaemia (Park 2015). Even after the birth of the child, the nutritional status of the mother influences that of the child through breastfeeding. Starting the pregnancy with a nutritional deficiency, lack of access to health services and clean water, repeated infections, and gut inflammation in mother and child after delivery can have an major impact on the absorption and utilisation of nutrients by them (UNICEF 2019).

3.3.4 Environmental-WaSH, Agro-climate

The immediate environmental conditions of an individual have an assertive consequence on their nutritional status. A clean hygienic environment, devoid of dirt accumulation and waterlogging, ensures better health and, in turn, a better absorption of nutrients and improved nutritional status.

Agro-climatic factors have also been studied as contributors to nutritional status. The geographical characteristics of an area, the type of agricultural land, and the rainfall in the area are closely linked to the capacity of food grain production, quality, and quantity of nutritious food available in a region and hence, the nutritional status of the people residing in that area. Regions with high disparity in landholdings and financial distribution, racial and cultural differences have shown greater disparities in the nutritional status of the people (Unisa et al. 2019).

3.3.5 MORBIDITIES

Undernutrition and infections are in vicious cycle, an inseparable cause–effect pair. Pathological conditions like diarrhoeal infections, fever, or malaria may hinder the absorption of nutrients by an individual. As a result, individuals become malnourished and deficient in vital vitamins and minerals, which in turn affects their immunity.

As a direct consequence of compromised body immunity, individuals now become susceptible to infections and infestations, thereby continuing the vicious cycle (Park 2015).

3.3.6 BEHAVIOURAL

The behavioural aspects of a community have greater implications for health status. The food habits of a community or a region have a deep-rooted cultural and psychological association with love, care, social acceptability, self-image, and social prestige. The diet of the people is influenced immensely, as discussed earlier, by the local conditions such as geography, climatic factors, religious customs, and beliefs. Against this backdrop, accessibility or the choice of food groups is often countered when individuals are faced with diseases that may result in severe deficiencies of both macro- and micronutrients.

3.3.7 POLITICS AND GOVERNANCE

Governance has a major role in the nutrition status of citizens, which it facilitates through nutrition policies, supplementary nutrition programmes, deficiency disorder prevention programmes, food and nutrition security nets, export–import policies, etc.

Nutrition policies often prescribe targeted reductions in major nutritional disorders of public health importance with locally appropriate interventions. Most countries have public distribution systems and other food security safety nets to ensure the availability of food to most of their citizens. Also, the export–import policies of the government with regards to foods are often regulated to keep up with the country's balance sheet.

Supplementary nutrition policies like the Integrated Child Development Services (ICDS), midday meal scheme, etc, have been successful models that have not only contributed to increased consumption patterns but also to nutrition monitoring and surveillance which in turn helps to strive for a better nutritional status, especially of vulnerable groups (Park 2015).

An important corollary to nutritional policies are the government interventions to improve accessibility to food by improving the quality and availability of infrastructure such as roads, ports, communication, and food storage facilities, and other resources that enhance the functioning of markets (Ekholuenetale 2020).

3.4 CONSEQUENCES OF UNDERNUTRITION AND HIDDEN HUNGER

The detrimental effects of undernutrition and hidden hunger span various dimensions such as physical, psychological (Hickson and Julian 2018), maternal, and socioeconomic aspects. Each of these aspects could be discussed as a function of time-dependent effects, i.e., immediate and delayed effects of undernutrition and hidden hunger.

3.4.1 PHYSICAL

The immediate consequence of deficiency of macro- and micronutrients can be observed in the weight of an individual. In children, undernutrition often results from an acute or chronic illness and related restricted diet, insufficient food intake, or poor appetite can lead to weight loss, lack of weight gain, or failure to thrive (American Academy of Pediatrics, Committee on Nutrition

2003). In adults, there is a loss of appetite and loss of muscle mass, which in turn impairs the muscle functions that are integral to the function of several major organs such as the heart, liver, and kidneys. For example, Carr et al. (1989) reported that undernutrition had aggravated congestive heart failure due to a heightened risk of tricuspid regurgitation and right atrial pressure (Carr et al. 1989).

Nutritional status, infection, and mortality by and large, have a strong correlation with nonspecific and specific immunity beyond the first of line of defence of the body, i.e., physical barriers including skin and mucous membranes. Deficiency of nutrients can compromise these natural barriers, increasing the susceptibility and risk of infections (Correia and Waitzberg 2003).

Micronutrients like iron, zinc, magnesium, copper, vitamins A, C, K, etc., are involved in the wound-healing processes such as clotting, blood vessel and tissue repair, fibroblast proliferation, and collagen synthesis. Deficiency of these micronutrients, as seen in undernourished individuals, may delay the wound-healing process (Stechmiller 2010).

Iodine deficiencies are well known in hilly areas and foothills. Much of the Indian research on iodine deficiencies was done in the Kangra Valley of Himachal Pradesh. Iodine is an important component of thyroid hormone, so deficiency of iodine causes increased thyroid stimulation through thyroid stimulating hormone (TSH), leading to global enlargement of the thyroid gland. This enlargement of the thyroid gland is called goitre. Females, especially during pregnancy, are more likely to be affected. Iodine deficiencies in children cause cretinism. Cretins have low a IQ and are mentally challenged.

Iron deficiency manifests most commonly as anaemia, since iron is an indispensable part of haemoglobin, the oxygen-carrying protein of the red blood cells (RBCs). Iron deficiency anaemia (IDA) is one of the most common dietary deficiency manifestations in the world, and more than half of Indian women are anaemic. Anemia causes weakness, decreased scholastic performance, post-partum haemorrhage, low birth weight, and can lead to heart failure.

Folic acid deficiencies are known to cause neural tube defects among foetuses when mothers are deficient during pregnancy. They also cause macrocytic anaemias like cyanocobalamin deficiency.

Vitamin A deficiency manifests as a spectrum of signs and symptoms. It is more common among children. It starts as night blindness as retinol is necessary for the corneal rod cells to function. Vitamin A is also essential for epithelialisation, so deficiency leads to conjunctival and corneal xerosis, which in the long term leads to corneal ulcers and to melting of the cornea, called as keratomalacia.

Vitamin D deficiency leads to rickets in children, generally in young children under 2 years, after weaning from the breast. Rickets is manifested as growth failure, hypocalcaemia, convulsions, and bone deformities like rickety rosary and kyphosis/scoliosis. Osteomalacia is the adult deficiency form, which is seen generally in women during pregnancy and lactation.

Iodine deficiency has multiple adverse effects on growth and development due to inadequate thyroid hormone production that are called iodine deficiency disorders. Iodine deficiency in pregnant mothers impairs motor and mental development of the foetus as well as increases the risk of miscarriages (Black et al. 2008).

3.4.2 Psychological

The functional consequences of such malnutrition include not only physical changes but also psychological changes such as depression, anxiety, irritability, apathy, poor sleep patterns, and loss of concentration (Stanga 2007). Malnutrition in young children may impair brain development by causing changes in the temporal sequence of brain, thereby disturbing the neural circuit formation (Udani 1992). Furthermore, the formation of essential brain structures is greatly impacted in malnourished children in the first three years of life compared to the brains of well-nourished children, resulting in irreversible changes. These children are at increased risk of cognitive deficits and learning disabilities (Kakietek 2017; Kar 2008). Additionally, undernourished children are more susceptible to delayed admissions to school, repeating grades, or dropping out of school (Mendez and Adair 1999).

3.4.3 SOCIOECONOMIC

Undernutrition levies a heavy cost either directly or indirectly on all the institutions of a society, be it an individual, family, or a nation. A tremendous loss in productivity, combined with an increased investment in healthcare, causes a decline in economic growth as well as an unstable human resource pool, in addition to child mortality, premature adult morbidities and mortalities that are linked to non-communicable diseases (Popkin et al. 2020; Nugent et al. 2020).

The earning potential of undernourished children as adults is estimated to be reduced by at least 10% of their overall lifetime earnings. This, in turn, can have implications on poverty, human capital, and gross domestic product (GDP) of the country (Hoddinott et al. 2008).

3.5 MANAGEMENT OF UNDERNUTRITION AND HIDDEN HUNGER

Management of undernutrition and hidden hunger depends on dealing with the underlying causes. The various risk determinants, such as behavioural, diseases, and social factors, influence the availability of optimum food and nutrients and need to be tackled in order to prevent and treat undernutrition and hidden hunger. The measures to manage undernutrition and hidden hunger are as follows:

3.5.1 IMPROVING DIETARY INTAKE

This is one of the direct approaches to manage undernutrition and hidden hunger. From a nutritional point of view, the prenatal, infancy, and adolescence periods are most sensitive, and hence during these stages, the increased nutritional demand has to be fulfilled. Food and nutrients supplementation during these stages should be ensured.

Although various supplementation programmes are running, their regularity and outreach to the target group need to be monitored. For instance, the ICDS scheme is one of the world's largest programmes for early childhood care and development, running in India since 1975 to improve the health status of children, pregnant women, and nursing mothers. Supplementary nutrition is one of its objectives; however, irregularity in ration supplies, lack of dietary diversity, insufficient funds, and poor basic infrastructure (Institute of Economic Growth 2020) have impeded its progress. Similarly, the PM POSHAN scheme is one of the world's largest school meal programmes functioning in India to address hunger and malnutrition among school-going children aged from 6 to 14 years. One hot cooked meal is provided to children attending school to address hunger, as many of them come to school with an empty stomach. Though there are strict regulations regarding the maintenance of nutritional standards and quality of meals, sometimes corrupt practices and poor hygiene conditions create hurdles to eradicating hunger and malnutrition. To overcome the abovementioned issues, development of low-cost yet diverse nutritious food products from local produce and ensuring their consumption by the beneficiaries is important. Improving basic infrastructure and ensuring 100% coverage of the target group under the nutrition programmes have to be guaranteed. Furthermore, regular monitoring of such programmes by the concerned officials should be made mandatory with stringent actions against wrongdoers.

Government-run eateries/canteens may be operated in all the states where at least a single balanced meal per day at a minimum charge could be made available to the poor, so that they may not go to bed hungry. It is important that local produce should be explored to prepare the meals and for the distribution of grains under the public distribution system. This approach can work multi-directionally in combating undernutrition and hunger by helping farmers to produce and earn, creating employment for local people in processing and cooking units, and making food available at a low cost for the hungry. This would address the issues of availability, accessibility, and affordability to some extent. Government, health workers, and nutritionists have to come together and work jointly in this arena.

In addition, exclusive breastfeeding until 6 months of age, improving the affordability of nutritious weaning foods by lowering the cost of weaning foods, increasing the availability of cost-effective fortified food products, and increasing the cost of junk food by levying taxes should be promoted to fight against hunger and malnutrition.

3.5.2 Reducing Morbidities

Undernutrition is not necessarily always a result of inadequate food intake. The quality of food and its proper digestion and absorption of nutrients in the body also impact nutritional state. The diseased state of a person hampers the absorption of essential nutrients, thereby leading to an immune-deficient state. The immune system compromised at the cost of poor nutritional state invites a host of infections in the body. These infections, when not treated because of a lack of efficient health services, may further aggravate the undernourished state and increase the risk of mortality. Poor sanitary conditions are associated with the transmission of infectious diseases, further exacerbating immunity and the undernourished state. There is an interrelationship between the underlying causes of the diseased state leading to undernutrition. Unhygienic conditions perpetuate infections, which reduce body immunity leading to diseases that, with an absence of healthcare facilities, leave the diseases untreated, eventually leading to undernutrition. Thus, to manage undernutrition and hidden hunger, availability and accessibility to healthcare are crucial. Immunisation is of prime importance. Awareness on the part of people and provision of health facilities for all on the part of the government have to be prioritised. There is a need to strengthen primary healthcare centres in terms of the availability of health professionals and health services. In case of a shortage of healthcare facilities, promoting telemedicine for remote and inaccessible areas would be of some respite for vulnerable populations. The rising cost of medicines and diagnosis needs to be curbed. Health assurance schemes should be provided by the governments for health coverage of poor and vulnerable families. Investing in preventive medicine should be encouraged. The revival of traditional and complementary medicine should be promoted as it would be able to fill the gap created by the shortage of healthcare infrastructure, health professionals, diagnostic facilities, and costly medicines.

3.5.3 Addressing Sociobehavioural Risk Factors

Addressing undernutrition and hidden hunger is as much an individual responsibility as it is the state's responsibility. Certain wrong social and cultural beliefs that interfere with the consumption of diverse foods should be stopped from passing to the next generation. For instance, discarding the colostrum-rich first breastmilk; giving prelacteal feeds like honey, water, and *ghutti* to babies before 6 months of age; delayed initiation of complementary feeding; restricting certain fruits, legumes, and vegetables to lactating and pregnant mothers, etc. These myths and taboos have adverse effects on the availability of nutrients, reduce immunity in infants, and affect their physical and cognitive development.

Gender discrimination is yet another social factor which has a direct impact on the quantitative and qualitative consumption of food. In many societies, girls and women are considered to play second fiddle in the family and therefore do not receive an equal share of the food, and ironically this is done by women (mothers) themselves. Therefore, female education is the key to eradicating hidden hunger. Female education would also play a primary role in family planning and thereby controlling the alarming rise in population, which is another reason increasing the threat to the sustenance of resources and their availability to all.

To break the vicious cycle of disease and malnutrition, maintenance of basic hygiene and sanitation and access to clean water are crucial. Education about the importance of sanitation and clean water must be imparted at the school level so that it serves as a long-term investment in health. Besides, nutrition education should be given at the community level as an immediate measure.

Food loss and wastage should be controlled. As per the United Nations (2022b), globally 13% of food is lost and 17% is wasted, summing up to 30% of total food waste and loss. Almost one third of food produced being lost and wasted is surely a threat to food security and the sustainability of the food system, contributing to hunger by lowering the availability of food and inflating food prices. The habit of sensible cooking at the household level, avoiding hoarding of food products, and ordering appropriately at food outlets should be inculcated. Also, robust post-harvest technology and storage at the retail level are required. Governments can come up with some effective remedies in this arena. Also, the realisation of a sense of social responsibility of not wasting food at the household level, social functions, and food outlets is needed. Food banks can be created for the poor and hungry rather than wasting the food.

Household wealth is directly proportional to the affordability of food, and therefore governments should plan programmes to provide work for everyone. The rate of unemployment has to be brought down. Creating job opportunities, providing education to all, and imparting vocational skills at the very beginning of life may aid as a long-term measure in improving household wealth. Programmes like Food for Work as an immediate measure for fighting against hunger should be effectively conducted.

3.5.4 ROLE OF GOVERNMENTS AND POLICYMAKERS

Governments and policymakers play a vital role in the eradication of undernutrition and hidden hunger. Worldwide, various international and national organisations are successful in the fields of health and nutrition, consequently benefiting a large segment of the population. However, sometimes sudden natural and man-made disasters impede progress. Conflicts within countries and wars across international borders are avoidable situations. International organisations can intervene in such avoidable situations to curtail the risk in terms of losses in food production, disruption of food supplies, displacement of affected populations, spreading of infections, loss of livelihoods, etc. Preparedness for natural disasters like floods, droughts, or sudden pandemics like COVID-19 should be one of the top priorities for governments. Immediate measures for providing humanitarian relief should be the main concern.

Agricultural production is fundamental to food availability, and farmers play a central role in food production. Therefore, governments must support farmers by providing subsidies for farm equipment, seeds, and fertilisers; extension education and services; storage, market, and minimum price support for their produce. Also, preparedness for dealing with climatic shocks should be amongst the priorities in agriculture policy.

Governments and policymakers should make robust food and nutrition security policies. Successful models of programmes running in the field of combating hidden hunger and undernutrition should be adopted and modified as per their sociodemographic needs.

Imparting nutrition education can act as a long-term measure to bring permanent change in the behaviour and attitude of people, and therefore governments should organise campaigns to create awareness among the those in need. A strong surveillance system should be deployed, and genuine feedback should be taken so as to bring about positive changes in the model. Political willpower is of utmost importance for the eradication of undernutrition and hidden hunger.

3.6 CONCLUSION

Undernutrition and hidden hunger are one of the biggest health challenges in the current era. A host of causes at the individual level, household level, and governance level is responsible for undernutrition and hidden hunger, affecting individuals physically, psychologically, and socioeconomically.

However, considering the management of undernutrition and hidden hunger as a social responsibility for each individual, strong political willpower will help in the eradication of undernutrition and hunger. Measures to improve availability, accessibility, and affordability of optimum yet nutritious food for all need to be adopted.

REFERENCES

American Academy of Pediatrics, Committee on Nutrition. 2003. Prevention of pediatric overweight and obesity. *Pediatrics* 112: 424–427.

Black, R.E., L.H. Allen, Z.A. Bhutta, L.E. Caulfield, M. De Onis, M. Ezzati, C. Mathers, J. Rivera, and Maternal and Child Undernutrition Study Group. 2008. Maternal and child undernutrition: Global and regional exposures and health consequences. *The Lancet* 371: 243–260.

Burgess, A. 2008. Undernutrition in adults and children: Causes, consequences and what we can do. *South Sudan Medical Journal* 1, no. 2: 18–22.

Carr, J.G., L.W. Stevenson, J.A. Walden, and D. Heber. 1989. Prevalence and hemodynamic correlates of malnutrition in severe congestive heart failure secondary to ischemic or idiopathic dilated cardiomyopathy. *American Journal of Cardiology* 63, no. 11: 709–713.

Correia, M.I., and D.L. Waitzberg. 2003. The impact of malnutrition on morbidity, mortality, length of hospital stay and costs evaluated through a multivariate model analysis. *Clinical Nutrition* 22, no. 3: 235–239.

Ekholuenetale, M., G. Tudeme, A. Onikan, and C.E. Ekholuenetale. 2020. Socioeconomic inequalities in hidden hunger, undernutrition, and overweight among under-five children in 35 sub-Saharan Africa countries. *Journal of the Egyptian Public Health Association* 95, no. 1: 1–15.

Hickson, M. and A. Julian. 2018. Consequences of undernutrition. *Advanced Nutrition and Dietetics in Nutrition Support*: 33–41. https://doi.org/10.1002/9781118993880.ch1.5

Hoddinott, J., J.A. Maluccio, J.R. Behrman, et al. 2008. Effect of a nutrition intervention during early childhood on economic productivity in Guatemalan adults. *The Lancet* 371: 411–416.

Institute of Economic Growth. 2020. Evaluation of ICDS scheme of India. Institute of Economic Growth, New Delhi. https://www.niti.gov.in/sites/default/files/2023-03

Kakietek, J., J.D. Eberwein, D. Walters, et al. 2017. *Unleashing Gains in Economic Productivity with Investments in Nutrition*. Washington, DC: World Bank Group.

Kar, B.R., S.L. Rao, and B.A. Chandramouli. 2008. Cognitive development in children with chronic protein energy malnutrition. *Behavioral and Brain Functions* 4, no. 1: 1–12.

Keeley, B., C. Little, and E. Zuehlke. 2019. *The State of the World's Children 2019: Children, Food and Nutrition – Growing Well in a Changing World*. UNICEF.

Mendez, M.A. and L.S. Adair. 1999. Severity and timing of stunting in the first two years of life affect performance on cognitive tests in late childhood. *The Journal of Nutrition* 129, no. 8: 1555–1562.

Nugent, R., C. Levin, J. Hale, et al. 2020. Economic effects of double burden of malnutrition. *The Lancet* 395: 156–164.

Park. 2015. *Park textbook of Preventive & Social Medicine*. 25th edition. Jabalpur: Banarsidas Bhanot Publishers.

Popkin, B.M., C. Corvalan, and L.M. Grummer-Strawn. 2020. Dynamics of the double burden of malnutrition and the changing nutrition reality. *The Lancet* 395, no. 10217: 65–74.

Stanga, Z., J. Field, S. Iff, et al. 2007. The effect of nutritional management on the mood of malnourished patients. *Clinical Nutrition* 26, no. 3: 379–382.

Stechmiller, J.K. 2010. Understanding the role of nutrition and wound healing. *Nutrition in Clinical Practice* 25, no. 1: 61–68.

Udani, P.M. 1992. Protein energy malnutrition (PEM), brain and various facets of child development. *The Indian Journal of Pediatrics* 59, no. 2: 165–186.

UNICEF. 2019. *The State of the World's Children 2019. Children, Food and Nutrition: Growing Well in a Changing World*. New York: UNICEF.

UNICEF/WHO/World Bank Group Joint Child Malnutrition Estimates. 2023. *United Nations Children's Fund (UNICEF), World Health Orgnization (WHO), International Bank for Reconstruction and Development/The World Bank. Levels and Tends in Child Malnutrition: UNICEF/WHO/World Bank Group Joint Child Malnutrition Estimates: Key Findings of the 2023 Edition*. New York: UNICEF and WHO.

Unisa, S., T. Lakhan, A. Saraswat, et al. 2019. Exploring the variations in adult height of population by agro-climatic zones in India. *The Journal of Family Welfare* 64: 18–26.

United Nation. 2022a. The sustainable development goal report 2022. https://unstats.un.org/sdgs/report/2022

United Nations. 2022b. Reducing food loss and waste: Taking action to transform food systems. https://www.un.org/en/observances/end-food-waste-day

WHO. 2022. UN Report: Global hunger number rose to as many as 821 million in 2021. https://who.int/news/item/06-07-22

4 Childhood Obesity

4.1 INTRODUCTION

The nutrition-related epidemiological transition around the globe has been taking place at a tremendous speed, influenced by urbanisation, globalisation, and the concomitant rise in income and purchasing power. This has resulted in a significant change in the quality and quantity of diet (WHO 2017a). A direct impact of dietary changes has been on the manifestation of malnutrition as well as nutrition-related chronic diseases (NRCDs) such as obesity, diabetes, cardiovascular diseases, and cancer (Sahoo et al. 2015).

The double burden of malnutrition has presented itself as one of the greatest global challenges affecting developed as well as developing nations alike. The double burden of malnutrition is characterised by the coexistence of two extreme forms of malnutrition: undernutrition and overnutrition. Undernutrition has various forms such as micronutrient deficiencies, being underweight, childhood stunting, and wasting whereas overnutrition encompasses overweight/obesity and diet-related non-communicable diseases (NCDs) (WHO 2017a; Popkin et al. 2020). For several decades, overnutrition was considered to be a problem characteristic of only developed nations. However, in recent times, the trend has spread rapidly to low- and middle-income nations as well (Ranjani et al. 2016; Kolcic 2012).

The condition of being overweight encompasses people with weight higher than what is considered healthy for a given height, while obesity is a measure of excess body fat. There has been an exponential rise in the obesity pandemic in the world. In 2016, more than 1.9 billion adults (>18 years) were overweight, of which 65% were obese. A large proportion of children and adolescents aged 5–19 years (>340 million children) were overweight or obese (WHO 2021).

Childhood obesity has become a serious public health challenge today. The significant consequence of childhood obesity is the fact that it makes its way into adulthood as well, along with other NCDs such as dyslipidaemia, hyperinsulinaemia, type-2 diabetes, hypertension, cardiovascular diseases, arthritis, and behavioural problems (Freedman et al. 1999; Laxmaiah et al. 2007).

With such rampant spread, it is important to monitor and better understand the different complex risk factors, consequences, and the prevention approaches, in order to have better-targeted interventions and policies in place for the management of childhood obesity. This chapter aims at discussing these various facets of childhood obesity in detail.

4.2 CHILDHOOD OBESITY

4.2.1 Definition

Overweight and obesity are defined by WHO (2021) as abnormal or excessive fat accumulation, and body mass index (BMI) is commonly used as a tool to classify adults as overweight and obese. BMI is calculated as the weight of a person in kilograms divided by the square of their height in metres (BMI=kg/m^2).

It is important to remember that BMI is not an actual measure of adiposity and can tend to over-estimate, particularly in the case of tall and lean children (Tyson and Frank 2018; NICE Guideline 2006). However, for a population-level estimation of overweight and obesity in adults, BMI is a very useful tool as it is the same irrespective of age and gender. For children though, age is an important factor.

DOI: 10.1201/9781003354024-4

According to WHO (2021), for children under 5 years of age: 'overweight is weight-for-height greater than 2 standard deviations above WHO Child Growth Standards median; and obesity is weight-for-height greater than 3 standard deviations above the WHO Child Growth Standards median'.

Similarly, for 5–19 years: 'overweight is BMI-for-age greater than 1 standard deviation above the WHO Growth Reference median; and obesity is greater than 2 standard deviations above the WHO Growth Reference median'.

4.3   EPIDEMIOLOGY OF CHILDHOOD OBESITY

The obesity pandemic is a result of composite factors such as changing lifestyles coupled with urbanisation and industrialisation.

The prevalence of overweight and obesity in children below the age of 5 years was published in April 2019 by UNICEF, WHO, and the World Bank (WHO 2019). According to this report, the prevalence of overweight individuals increased from 4.8% in 1990 to 5.9% or approximately 38.3 million children in 2018, however, the trends were found to be heterogeneous in low- and middle-income countries. Though close to half of the estimated overweight children were living in Asia, Africa constituted another quarter, the condition was a concern for all the regions of the world. A significant increase in the number of overweight and obese children under 5 years of age was observed in the case of South-Eastern and North American United Nations regions.

The report further stated that from 2000 to 2019, there was no progress in any of the countries in any income group that could stem the rate. Some 8% of all overweight children under 5 live in low-income countries while 37% belonged to lower-middle-income countries. The highest percentage of prevalence was however from upper-middle-income countries where 41% of children belonged to the overweight category.

The estimates of overweight and obesity in children aged 5–19 years were available from the NCD-Risk factor Collaboration (NCD-RisC), which has the largest global database, with the most recent estimates available from 2016 (NCD-RisC 2017). The estimates reported by NCD-RisC showed that the prevalence of obesity had increased from 0.7% to 5.6% in girls while it was a jump from 0.9% to 7.8% in boys of 5–19 years. These estimates varied from those reported by the Global Burden of Disease (GBD) study conducted in 2013 which had shown that the prevalence of obesity among children and adolescent boys worldwide was 12.9% and 13.4% for girls (Ng et al. 2014).

The transition from undernutrition to overnutrition in lower- and middle-income countries is a grave concern which has led to the double burden of malnutrition, which can be attributed to the unequal economic and demographic transitions (Prentice 2018). This observation is evident from the estimates of Brazil, where the percentage prevalence of overweight children aged 6–18 years of age had drastically increased from 4.1% to 13.9%, whereas there was a dip in the prevalence of underweight from 14.8% to 8.6%. Another similar concern was observed in Mozambique, Africa, where the prevalence of overweight had increased in children below 5 years of age from 3.6% to 7.9% during the period 2005–2014, while stunting had remained constant at 43.7% and 43.1% (Wang and Lobstein 2006; Wang et al. 2002).

The estimates of overweight and obese children aged 0–19 years of age in India have been recorded by the Comprehensive National Nutrition Survey (CNNS), 2019. According to the CNNS report (CNNS 2019), 2% of children under 5 years of age were overweight or obese as per the standards (WHZ >+2 SD). The National Family Health Survey (NFHS-5), carried out in 2019–2021, has reported a prevalence of 3.4% overweight and obese children in India (NFHS-5 2021). The prevalence of overweight or obesity in school-going children as recorded in CNNS was found to be 4% (BMI-for-age >+1 SD). The rates of overweight and obesity have been found to be gradually increasing with age, with 5% of adolescents recorded to be overweight or obese (BMI-for-age >+1 SD).

4.4   RISK FACTORS

Childhood obesity is an interplay of multiple factors, both modifiable and non-modifiable risk factors. These factors have an interdependent cause–effect relationship with childhood obesity (Figure 4.1).

4.4.1   Genetic Factors

Obesity is greatly influenced by genetic factors. The chances of an individual becoming obese are higher when relatives are obese. This hypothesis has been studied on both nuclear and extended families as well as on twins and siblings (Kurpa and Swaminathan 2021). Gene–environment interactions are also equally responsible for the manifestation of overweight or obesity in children at later stages. Gene–environment interactions have shown to cause epigenetic modifications. These include the physiological adaptations and foetal programming that occur due to reduced nutrient availability during the intrauterine period. These adaptations are often life-long (Loke et al. 2008).

Ethnicity is another important factor that influences the incidence of obesity. Whincup et al. (2002) studied the ethnic differences in cardiovascular risk and reported an increased sensitivity of British Asian children to adiposity as compared to other British children.

4.4.2   Maternal Factors

Childhood obesity is positively linked to maternal factors such as maternal obesity, gestational weight gain, and glycaemia during pregnancy (Catalano and Ehrenberg 2006; Hillier et al. 2007). In addition, maternal undernutrition during gestation is also deeply associated with childhood undernutrition and obesity. The Dutch Famine Birth Cohort study and the Great Chinese Famine indicated that women who were exposed to the famine in early gestation showed higher rates of overweight and obesity (Ravelli et al. 1999; Wang et al. 2010). On the contrary, mothers consuming a high glycaemic index diet during pregnancy have been associated with large-for-gestational

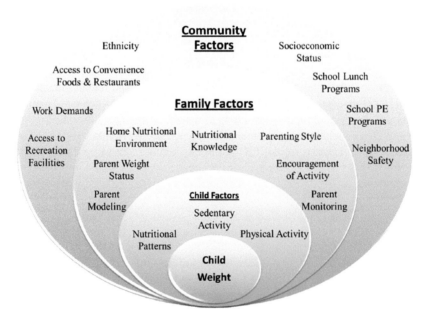

FIGURE 4.1 Ecological model of childhood obesity. (Adapted from Davison, K.K., Birch, L.L., Childhood overweight: a contextual model and recommendations for future research. *Obesity Reviews*. 2001, 2(3), 159–171. With permission.)

age children (Moses et al. 2006). A similar correlation has been observed by Chen et al. (2017) in their 'Growing Up in Singapore Towards Healthy Outcomes' or GUSTO study. The study shows a higher infancy and early childhood BMI associated with higher carbohydrate and sugar intake by the pregnant mothers. Prenatal maternal lifestyle is a major indicator of offspring adiposity. It has been observed that prenatal maternal smoking has a consistent association with the risk of children being overweight or obese (Khan et al. 2017; Oken et al. 2008; Li et al. 2016).

4.4.3 DIETARY FACTORS

Breastfeeding has been associated with lowering the risk of obesity (Victoria et al. 2016). Compared with formula-fed infants, the growth pattern of breastfed infants is slower. Additionally, the nutrient composition of formula feed is higher compared to breastmilk. Breastfed children have been found to have better insulin levels (Kurpa and Swaminathan 2021; De Onis et al. 2007). Initiation of complementary feeding before the age of 6 months has been associated with a higher risk of obesity compared to children who were given complementary feeding only after 6 months of exclusive breastfeeding. Dietary patterns established during the complementary feeding stage have been observed to persist even during later years of life (Lanigan et al. 2019).

During the preschool age, it has been noted that most of the dietary recommendations are not met. Consumption of energy-dense foods during childhood has been implicated in weight gain and obesity. Energy-dense foods have been studied to reduce satiety, thereby causing overeating in adolescents (Krebs et al. 2007). The Avon Longitudinal Study of Parents and Children (ALSPAC) study revealed that consumption of energy-dense, low-fibre, and high fat foods at ages 5–7 years was associated with a greater chance of adiposity at 9 years of age (Johnson et al. 2008).

4.4.4 PHYSICAL ACTIVITY

Exercise and physical activity form an integral part of growth and development during the early stages of life. Physical activity, coupled with good nutrition, can ensure optimal growth of a child while also lowering the probability of obesity. In the last two decades, the shift in the lifestyles of the population to an increased sedentary life and physical inactivity has been studied to impact the incidence of obesity to a great extent. In a study conducted by Laxmaiah et al. (2007) in Hyderabad, India, the effect of physical activity on the occurrence of obesity and overweight in adolescents was observed. The study reported a strong correlation between the prevalence of overweight/obesity and physical activity in adolescents in terms of participation in outdoor games, exercise, household chores, or watching television.

An important aspect of physical activity is the transition in life stages. During the transition from preschool to school-age and from school-age to adolescence, there are several biological transformations that happen to the body which may particularly influence the physical activity of an individual, which in turn affects body weight. A classic case is when a child hits puberty and the changes in body composition make children self-conscious, thereby reducing their willingness to engage in physical activity. This invariably results in an energy imbalance, leading to obesity in subsequent years (Venn et al. 2007).

4.4.5 ENVIRONMENTAL FACTORS

The environment in which the children grow has a major impact on their predisposition to obesity. Socioeconomic status, for instance, has been studied to influence the occurrence of overweight in the children from lower socioeconomic status, as reported in the European Young Heart Study (Kristensen et al. 2006). In India, however, obesity and overweight are seen to be more prevalent in children belonging to the higher socioeconomic status group (Ramachandra et al. 2002; Bhardwaj et al. 2008).

Another important contributor to the global obesity pandemic is rapid urbanisation, which has led to improved availability and accessibility of meat, butter, oils, fats, and sugars on the one hand, and increasing the prices of fruit and vegetables on the other. This, in turn, has severely influenced the adiposity of the populations (James 2008). Changing food prices, increased advertising promotional activities via mass media, and improved liberal agricultural policies are some of the key influencing factors in creating an 'obesogenic environment' (Popkin 2006; St-Onge et al. 2003).

4.5 CONSEQUENCES OF CHILDHOOD OVERWEIGHT AND OBESITY

The manifestation of childhood obesity can be evaluated under the following headings:

4.5.1 MEDICAL CONSEQUENCES

The consequences of childhood obesity can be broadly classified into medical and psychological consequences. Medical consequences include metabolic complications such as diabetes mellitus, hypertension, dyslipidaemia and non-alcoholic fatty liver disease and mechanical complications such as obstructive sleep apnoea syndrome and orthopaedic problems such as genu varus and valgus deformity of the knees, Blount's disease, and slipped capital femoral epiphysis (Lee 2009). The metabolic complications of childhood obesity often referred to as metabolic syndrome consists of insulin resistance, often seen β-cell failure, impaired glucose tolerance, diabetes, dislipidemia, hypertension and premature heart disease.

Several physiological changes and medical outcomes have been linked to childhood obesity, which were predominantly found in only adults until recently, such as fatty liver disease, sleep apnoea, type 2 diabetes, asthma, hepatic steatosis (fatty liver disease), cardiovascular disease, high cholesterol, cholelithiasis (gallstones), glucose intolerance and insulin resistance, skin conditions, menstrual abnormalities, impaired balance, and orthopaedic problems (Lakshman et al. 2012). These disease conditions are, however, preventable and can subside once a healthy weight is reached (Niehoff 2009).

4.5.2 PSYCHOSOCIAL CONSEQUENCES

The psychosocial consequences of childhood obesity appear more widespread and prominent than the medical consequences, with the primary concern being 'stigmatising' and 'least social acceptance' of the obese children (Schwimmer et al. 2003). Children who are obese face stigma and bias in various environments, from their parents to school, friends, clinical settings, and even researchers (Lydecker et al. 2018).

Overweight or obese children have been found to have a reduced health-related quality of life in socioemotional aspects. Obesity onset in childhood can cause poor body image, low self-esteem, and confidence in children since that period forms the critical age of mental and emotional development. In the absence of this, the personality development of the child is immensely impacted (Fallon et al. 2005; Monello and Mayer 1963; Stunkard and Burt 1967). This is particularly true for adolescent overweight girls, who slip into negative self-image which usually continues into their adulthood.

These children also experience low self-esteem and depressive symptoms compared to normal children (Muhlig et al. 2016). Family environment and health behaviours have a deep impact on the outcomes of childhood obesity. Other implicating factors include obesogenic risk factors. Depressive symptoms are largely associated with reduced physical activity, a sedentary lifestyle, and improper diet (Hoare et al. 2014). Sheinbein et al. (2019) have reported that obese children who display depressive symptoms are associated with causative factors such as child eating disorder pathology, emotionally manipulative parenting lifestyle, and lower child social status (Figure 4.2).

4.5.3 ECONOMIC CONSEQUENCES

The economic consequences of childhood obesity can be understood in the light of direct effects such as medical costs as well as indirect effects such as labour market costs (Cawley 2010). Trasande and Chatterjee (2009) reported an estimated medical cost of childhood obesity associated with hospitalisations, charges,

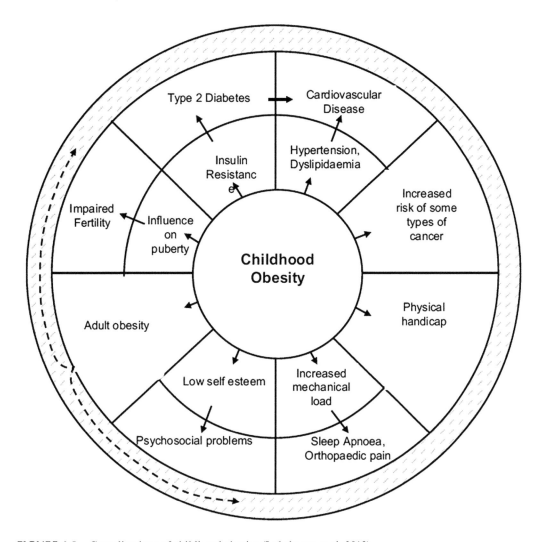

FIGURE 4.2   Complications of childhood obesity (Lakshman et al. 2012).

and costs of admissions to US hospitals to be around USD237.6 million in 2005. The costs rise much higher when obese children turn into obese adults, which is the case with about one-third of preschool children and about half of obese school-age children (Serdula et al. 1993). Obesity is associated with delayed skill acquisition in children and frequent job absenteeism in adults, thereby lowering productivity and constituting the indirect costs of obesity (Cawley and Speiss 2008; Gates et al. 2008).

4.6   PREVENTION AND CONTROL OF CHILDHOOD OBESITY

The World Health Organization has set a global target of no further increase in childhood overweight by 2025 (WHO 2017b). Childhood obesity prevention requires a holistic approach with intervention strategies aimed at multiple levels.

4.6.1   Prenatal and Neonatal Level

Optimum nutrition during pregnancy and after delivery is of utmost importance in tackling childhood obesity. Exclusive breastfeeding of infants up to 6 months of age and introduction of complementary feeding thereafter may help provide proper nutrition to the child. Government guidelines highlighting the importance of optimal nutrition to expectant and new mothers is an absolute necessity.

4.6.2 HOME LEVEL

Children should be involved in daily chores at home. Exercise and physical activity should be made a family affair where all family members are involved. It is important for parents to monitor screen time, eating behaviours, and sleeping patterns of their children to track their health status.

4.6.3 SCHOOL-BASED PROGRAMMES

School is an excellent place to start sensitisation in children of the importance of nutrition and physical activity, thereby preventing obesity. Ensuring balanced and healthy food in school cafeterias, banning sweetened beverages and junk food which is laden with calories and without any nutritional contribution, incorporation of knowledge on nutrition and physical activity in the school curriculum may help improve the nutritional status of the school-going children.

4.6.4 COMMUNITY OR NATIONAL LEVEL

At a larger scale, it is important to plan and execute programmes related to creating awareness about childhood obesity. Conducting workshops, organising health walks and healthy food festivals, and talking about balanced nutrition are some of the ways to introduce the concepts to a community in particular and the nation at large.

Governments should ensure that there is ample free space such as parks, playgrounds, and walking or cycling tracks available for citizens to choose from. It is also important to conduct periodic monitoring of the population for obese and overweight individuals and maintain a database of their nutritional statuses (Brown et al. 2015).

4.6.5 POLICY AND GUIDELINES

Setting up a national task force for obesity may be useful in ensuring continuous monitoring of obesity in the community. Food safety authorities may introduce stringent laws and guidelines with regards to promotional and marketing activities of the food industry. These guidelines specifically cater to the restriction on the advertisement of unhealthy and high fat-salt-sugar foods for children. Front-of-pack labelling should be introduced in all countries to provide an easy-to-understand warning label on the packaging whenever the commodities exceed the limits of calories, sugar, and saturated fats. This will facilitate consumers to make an informed choice (WHO 2018; Gupta et al. 2012).

4.7 CONCLUSION

Childhood obesity has become a serious public health challenge today. The significant consequence of childhood obesity is the fact that it makes its way into adulthood as well, along with other NCDs. Childhood obesity is at the interplay of modifiable and non-modifiable risk factors. These factors have an interdependent cause–effect relationship with childhood obesity. The repercussions of childhood obesity are reflected in terms of psycho-physiological health and economic losses. Childhood obesity prevention requires a holistic approach with intervention strategies targeted at multiple levels.

REFERENCES

Bhardwaj, S., A. Misra, L. Khurana, et al. 2008. Childhood obesity in Asian Indians: A burgeoning cause of insulin resistance, diabetes and sub-clinical inflammation. *Asia Pacific Journal of Clinical Nutrition* 17: 172–175.

Brown, C.L., E.E. Halvorson, G.M. Cohen, et al. 2015. Addressing childhood obesity: Opportunities for prevention. *Pediatric Clinics* 62, no. 5: 1241–1261.

Catalano, P.A., and H.M. Ehrenberg. 2006. The short-and long-term implications of maternal obesity on the mother and her offspring. *BJOG: An International Journal of Obstetrics & Gynaecology* 113, no. 10: 1126–1133.

Cawley, J. 2010. The economics of childhood obesity. *Health Affairs* 2, no. 93: 364–371.

Cawley, J., and C.K. Spiess. 2008. Obesity and skill attainment in early childhood. *Economics & Human Biology* 6, no. 3: 388–397.

Chen, L.W., I.M. Aris, J.Y. Bernard, et al. 2017. Associations of maternal macronutrient intake during pregnancy with infant BMI peak characteristics and childhood BMI. *The American Journal of Clinical Nutrition* 105, no. 3: 705–713.

CNNS. 2019. *Comprehensive National Nutrition Survey (CNNS) National Report.* Ministry of Health and Family Welfare (MoHFW), Government of India, UNICEF and Population Council, New Delhi.

De Onis, M., C. Garza, A.W. Onyango, et al. 2007. Comparison of the WHO child growth standards and the CDC 2000 growth charts. *The Journal of Nutrition* 137, no. 1: 144–148.

Fallon, E.M., M. Tanofsky-Kraff, A.C. Norman, et al. 2005. Health-related quality of life in overweight and nonoverweight black and white adolescents. *The Journal of Pediatrics* 147, no. 4: 443–450.

Freedman, D.S., W.H. Dietz, and S.R. Srinivasan. 1999. The relation of overweight to cardiovascular risk factors among children and adolescents: The Bogalusa Heart Study. *Pediatrics* 103, no. 6: 1175–1182.

Gates, D.M., P. Succop, B.J. Brehm, et al. 2008. Obesity and presenteeism: The impact of body mass index on workplace productivity. *Journal of Occupational and Environmental Medicine* 50: 39–45.

Gupta, N., K. Goel, P. Shah, et al. 2012. Childhood obesity in developing countries: Epidemiology, determinants, and prevention. *Endocrine Reviews* 33, no. 1: 48–70. https://doi.org/10.1210/er.2010-0028

Hillier, T.A., K.L. Pedula, and M.M. Schmidt. 2007. Childhood obesity and metabolic imprinting: The ongoing effects of maternal hyperglycemia. *Diabetes Care* 30, no. 9: 2287–2292.

Hoare, E., H. Skouteris, M. Fuller-Tyszkiewicz, et al. 2014. Associations between obesogenic risk factors and depression among adolescents: A systematic review. *Obesity Reviews* 15, no. 1: 40–51.

James, W.P.T. 2008. The epidemiology of obesity: The size of the problem. *Journal of Internal Medicine* 263, no. 4: 336–352.

Johnson, L., A.P. Mander, L.R. Jones, et al. 2008. Energy-dense, low-fiber, high-fat dietary pattern is associated with increased fatness in childhood. *The American Journal of Clinical Nutrition* 87, no. 4: 846–854.

Khan, M.N., M.M. Rahma, A.A. Shariff, et al. 2017. Maternal undernutrition and excessive body weight and risk of birth and health outcomes. *Archives of Public Health* 75, no. 1: 1–10.

Kolcic, I. 2012. Double burden of malnutrition: A silent driver of double burden of disease in low–and middle–income countries. *Journal of Global Health* 2, no. 2. https://doi.org/10.7189/jogh.02.020303

Krebs, N.F., J.H. Himes, D. Jacobson, et al. 2007. Assessment of child and adolescent overweight and obesity. *Pediatrics* 120, no. S4: S193–S228.

Kristensen, P.L., N. Wedderkopp, N.C. Moller, et al. 2006. Tracking and prevalence of cardiovascular disease risk factors across socio-economic classes: A longitudinal substudy of the European Youth Heart Study. *BMC Public Health* 6, no. 1: 1–9.

Kurpa, A., and S. Swaminathan. 2021. Prevention and management of overweight and obesity in children. In Vir, S.C. (Ed.), *Public Health Nutrition in Developing Countries*, 2nd Edition. Delhi: Woodhead Publishing India Pvt Ltd, 615–648. eBook ISBN 9780429091490.

Lakshman, R., C.E. Elks, and K.K. Ong. 2012. Childhood obesity. *Circulation* 126, no. 14: 1770–1779.

Lanigan, J., L. Tee, and R. Brandreth. 2019. Childhood obesity. *Medicine* 47, no. 3: 190–194.

Laxmaiah, A., B. Nagalla, K. Vijayaraghavan, et al. 2007. Factors affecting prevalence of overweight among 12-to 17-year-old urban adolescents in Hyderabad, India. *Obesity* 15, no. 6: 1384–1390.

Lee, Y.S. 2009. Consequences of childhood obesity. *Annals of the Academy of Medicine of Singapore* 38, no. 1: 75–77.

Li, L., H. Peters, A. Gama, et al. 2016. Maternal smoking in pregnancy association with childhood adiposity and blood pressure. *Pediatric Obesity* 11, no. 3: 202–209.

Loke, K.Y., J.B. Lin, and D.Y. Mabel. 2008. 3rd college of paediatrics and child health lecture—The past, the present and the shape of things to come. *Annals of the Academy of Medicine of Singapore* 37, no. 5: 429–34.

Lydecker, J.A., E. O'Brien, and C.M. Grilo. 2018. Parents have both implicit and explicit biases against children with obesity. *Journal of Behavioral Medicine* 41: 784–791.

Monello, L.F., and J. Mayer. 1963. Obese adolescent girls: An unrecognized "minority" group?. *The American Journal of Clinical Nutrition* 13, no. 1: 35–39.

Moses, R.G., M. Luebcke, W.S. Davis, et al. 2006. Effect of a low-glycemic-index diet during pregnancy on obstetric outcomes. *The American Journal of Clinical Nutrition* 84, no. 4: 807–812.

Muhlig, Y., J. Antel, M. Focker, et al. 2016. Are bidirectional associations of obesity and depression already apparent in childhood and adolescence as based on high-quality studies? A systematic review. *Obesity Reviews* 17, no. 3: 235–249.

NCD-RisC (NCD Risk Factor Collaboration). 2017. Worldwide trends in body-mass index, underweight, overweight, and obesity from 1975 to 2016: A pooled analysis of 2416 population-based measurement studies in 128.9 million children, adolescents, and adults. *Lancet* 390: 2627–2642.

NFHS-5. 2021. *National Family Health Survey-2019-21*. IIPS, Mumbai. https://main.mohfw.gov.in/sites/default/files/NFHS-5_Phase-II_0.pdf

Ng, M., T. Fleming, M. Robinson, et al. 2014. Global, regional, and national prevalence of overweight and obesity in children and adults during 1980–2013: A systematic analysis for the Global Burden of Disease Study 2013. *The Lancet* 384, no. 9945: 766–781.

NICE (National Institute for Health and Clinical Excellence, Great Britain). 2006. *Obesity: The Prevention, Identification, Assessment and Management of Overweight and Obesity in Adults and Children: NICE Guideline*. NICE.

Niehoff, V. 2009. Childhood obesity: A call to action. *Bariatric Nursing and Surgical Patient Care* 4, no. 1: 17–23.

Oken, E., E.W. Levitan, and M.W. Gillman. 2008. Maternal smoking during pregnancy and child overweight: Systematic review and meta-analysis. *International Journal of Obesity* 32, no. 2: 201–210.

Popkin, B.M. 2006. Global nutrition dynamics: The world is shifting rapidly toward a diet linked with non communicable diseases. *The American Journal of Clinical Nutrition* 84, no. 2: 289–298.

Popkin, B.M., C. Corvalan, and L.M. Grummer-Strawn. 2020. Dynamics of the double burden of malnutrition and the changing nutrition reality. *The Lancet* 395, no. 10217: 65–74.

Prentice, A.M. 2018. The double burden of malnutrition in countries passing through the economic transition. *Annals of Nutrition and Metabolism* 72, no. 3: 47–54.

Ramachandran, A., C. Snehalatha, R. Vinitha, et al. 2002. Prevalence of overweight in urban Indian adolescent school children. *Diabetes Research and Clinical Practice* 57, no. 3: 185–190.

Ranjani, H., T.S. Mehreen, R. Pradeepa, et al. 2016. Epidemiology of childhood overweight & obesity in India: A systematic review. *The Indian Journal of Medical Research* 143, no. 2: 160.

Ravelli, A.C., J.H. Van Der Meulen, C. Osmond, et al. 1999. Obesity at the age of 50 y in men and women exposed to famine prenatally. *The American Journal of Clinical Nutrition* 70, no. 5: 811–816.

Sahoo, K., B. Sahoo, A.K. Choudhury, et al. 2015. Childhood obesity: Causes and consequences. *Journal of Family Medicine and Primary Care* 4, no. 2: 187.

Schwimmer, J.B., T.M. Burwinkle, and J.W. Varni. 2003. Health-related quality of life of severely obese children and adolescents. *JAMA* 289, no. 14: 1813–1819.

Serdula, M.K., D. Ivery, R.J. Coates, et al. 1993. Do obese children become obese adults? A review of the literature. *Preventive Medicine* 22, no. 2: 167–177.

Sheinbein, D.H., R.I. Stein, J.F. Hayes, et al. 2019. Factors associated with depression and anxiety symptoms among children seeking treatment for obesity: A social-ecological approach. *Pediatric Obesity* 14, no. 8: e12518.

St-Onge, M.P., K.L. Keller, and S.B. Heymsfield. 2003. Changes in childhood food consumption patterns: A cause for concern in light of increasing body weights. *The American Journal of Clinical Nutrition* 78, no. 6: 1068–1073.

Stunkard, A., and V. Burt. 1967. Obesity and the body image: II. Age at onset of disturbances in the body image. *American Journal of Psychiatry* 123, no. 11: 1443–1447.

Trasande, L., and S. Chatterjee. 2009. The impact of obesity on health service utilization and costs in childhood. *Obesity* 17, no. 9: 1749–1754. https://doi.org/10.1038/oby.2009.67

Tyson, N., and M. Frank. 2018. Childhood and adolescent obesity definitions as related to BMI, evaluation and management options. *Best Practice & Research Clinical Obstetrics & Gynaecology* 48: 158–164.

Venn, A.J., R.J. Thomson, M.D. Schmidt, et al. 2007. Overweight and obesity from childhood to adulthood: A follow-up of participants in the 1985 Australian Schools Health and Fitness Survey. *Medical Journal of Australia* 186, no. 9: 458–460.

Victoria, C.G., R. Bahl, A.J. Barros, et al. 2016. Breastfeeding in the 21st century: Epidemiology, mechanisms, and lifelong effect. *The Lancet* 387, no. 10017: 475–490.

Wang, Y., and T. Lobstein. 2006. Worldwide trends in childhood overweight and obesity. *International Journal of Pediatric Obesity* 1: 11–25.

Wang, Y., C. Monteiro, and B.M. Popkin. 2002. Trends of obesity and underweight in older children and adolescents in the United States, Brazil, China, and Russia. *American Journal of Clinical Nutrition* 75, no. 6: 971–977.

Wang, Y., X. Wang, Y. Kong, et al. 2010. The Great Chinese Famine leads to shorter and overweight females in Chongqing Chinese population after 50 years. *Obesity* 18, no. 3: 588–592.

Whincup, P.H., J.A. Gilg, O. Papacosta, et al. 2002. Early evidence of ethnic differences in cardiovascular risk: Cross sectional comparison of British South Asian and white children. *BMJ* 324, no. 7338: 635.

WHO. 2017a. The double burden of malnutrition. Policy brief. Geneva: World Health Organization. https://www.who.int/publications/i/item/who-nmh-nhd-17.3

WHO. 2017b. Global targets 2025 to improve maternal, infant and young children nutrition (No. WHO/NMH/NHD/17.10). World Health Organization.

WHO. 2018. Taking action on childhood obesity (No. WHO/NMH/PND/ECHO/18.1). World Health Organization.

WHO. 2019. Global database on child health and malnutrition. UNICEF-WHO-The World Bank: Joint child malnutrition estimates – Levels and trends. Geneva: WHO. http://www.who.int/nutgrowthdb/estimates/en/ (accessed 15 September 2022).

WHO. 2021. Overweight and obesity. https://www.who.int/news-room/fact-sheets/detail/obesity-and-overweight

5 Diabetes

5.1 INTRODUCTION

Diabetes mellitus is a metabolic disorder characterised by inappropriate sugar levels in the blood. It occurs basically due to inefficient or insufficient secretion of the hormone insulin, which is responsible for managing blood sugar levels. This prolonged condition may lead to damage of several vital organs, such as the eyes, kidney, heart, and blood vessels. The severe complications arising as a result of improper blood sugar management increase the risk of disability and death, thereby posing a major health challenge.

5.2 PREVALENCE

According to the WHO (2020), diabetes was the ninth leading cause of death in 2019, causing 1.5 million deaths. It is estimated that worldwide about 463 million people were diabetic in 2019, which will increase to 578 million by 2030 and 700 million by 2045. The prevalence is estimated to be higher in urban areas and high-income countries (Saeedi et al. 2019), making it a disorder of affluence.

Impaired glucose tolerance (IGT), prevalent in 374 million in 2019, is estimated to rise to 454 million by 2030 and 548 million by 2045 (Saeedi et al. 2019).

The global burden of diabetes has increased significantly in the past three decades. The global incidence and prevalence of diabetes has seen an increase of 102.9% and 129.7%, respectively, from 1990 to 2017. An increase of 125.5% in deaths due to diabetes was estimated during the period. Likewise, the disability adjusted life years (DALYs) soared from 31.3 million in 1990 to 67.9 million in 2017, i.e., an increase of 116.7% The geographic distribution of the diabetic burden in 2017 showed the highest prevalence of diabetes in China followed by India and the US; and highest death and DALYs in India followed by China and Indonesia (Lin et al. 2020).

5.3 CLASSIFICATION OF DIABETES

Diabetes mellitus (DM) is broadly classified into the following subtypes:

5.3.1 Type 1 Diabetes Mellitus (T1DM)

It is also known as insulin-dependent diabetes mellitus (IDDM). As the name indicates, this form of diabetes occurs when there is no or very low insulin secretion by the body. Insulin is a hormone secreted by β-cells of the endocrine pancreas and in the case of T1DM, autoimmune destruction of β-cells results in insulin deficiency, consequently increasing the levels of sugar in the blood. T1DM represents 10% of total diabetic cases and occurs at an early age (Paschou et al. 2018).

5.3.2 Type 2 Diabetes Mellitus (T2DM)

It is also known as non-insulin dependent diabetes mellitus (NIDDM) and as the name indicates in this form of diabetes, insulin is secreted but it is inefficient in controlling levels of blood sugar. It is caused due to impaired insulin secretion and the inability of insulin-sensitive tissues to respond appropriately to insulin (Galicia-Garcia et al. 2020). Impaired insulin secretion and insulin resistance jointly lead to the progression of T2DM (Kaku 2010).

 DOI: 10.1201/9781003354024-5

5.3.3 GESTATIONAL DIABETES MELLITUS (GDM)

It is a complication of pregnancy characterised by chronic hyperglycaemia. Owing to various physiological changes during pregnancy, the state of insulin sensitivity initially in pregnancy and insulin resistance later lead to β-cell dysfunction and consequently GDM. It has been reported that worldwide GDM affects approximately 16.5% of pregnancies (Plows et al. 2018).

Besides the above-mentioned DM, a few other rare subtypes of DM are maturity-onset diabetes of the young (MODY), neonatal diabetes, latent autoimmune diabetes in adults (LADA), type 3c diabetes, cystic fibrosis diabetes, viral infection, and medication-induced diabetes (Banday et al. 2020; Yau et al. 2021). The WHO (2019) has given the revised classification of diabetes shown in Figure 5.1. Since discussing all the subtypes in a chapter is implausible, this chapter will focus on the major forms of diabetes, i.e., T1DM, T2DM, and GDM.

5.4 SYMPTOMS AND SIGNS

The classic signs of diabetes are polyuria, polydypsia, polyphagia, unexpected weight loss, frequent fatigue, irritability, repeated infections, dry mouth, burning, pain, numbness in the feet, itching, acanthoses nigricans, decreased vision, impotence or erectile dysfunction (Ramachandran 2014). Abnormal thirst, frequent urination, weight loss, genital itching, stomatitis, visual disturbances, fatigue, confusion, and balanitis are the symptoms associated with hyperglycaemias in newly diagnosed T2DM (Drivsholm et al. 2005).

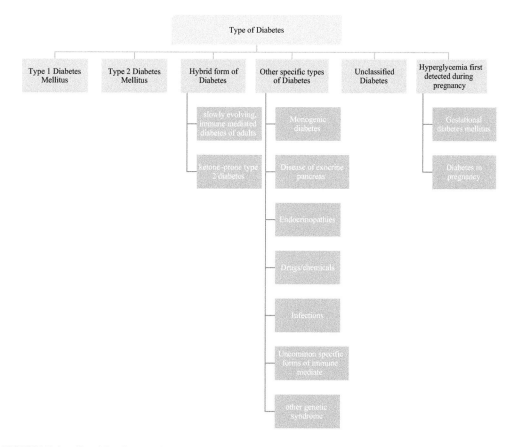

FIGURE 5.1 Classification of diabetes according to WHO.

5.5 DIAGNOSIS

An oral glucose tolerance test (OGTT) is performed to screen and diagnose individuals with prediabetes or diabetes. Fasting glucose in the range of ≥110 to ≤125 mg/dL is diagnosed as impaired fasting glucose (IFG), two-hour glucose in the range of ≥140 to <200 mg/dL is diagnosed as impaired glucose tolerance (IGT). Glycosylated haemoglobin in the range of 5.7–6.4% indicates prediabetes (Ramachandran 2014).

Fasting glucose ≥126 mg/dL, two-hour glucose ≥200 mg/dL, and glycosylated haemoglobin ≥6.5% is diagnosed as diabetes (Ramachandran 2014).

In T1DM, the presence of islet autoantibodies is used as a diagnostic criterion in the initial stage which is characterised by autoimmunity, normoglycaemia, and presymptomatic. The diagnostic criteria for the second stage of T1DM are the presence of islet autoantibodies and IFG and/or IGT, which is characterised by autoimmunity, dysglycaemia, and presymptomatic. The third stage is diagnosed by standard criteria for diagnosing diabetes, and autoantibodies may be absent. This stage is characterised by autoimmunity, overt hyperglycaemia, and symptomatic (American Diabetes Association Professional Practice Committee 2022).

Under asymptomatic conditions, the screening criteria for prediabetes and T2DM in adults are: 1) overweight or obese with one or more risk factors such as first-degree relatives with diabetes, high-risk race/ethnicity, acanthosis nigricans, cardiovascular disease, etc., 2) women with GDM, 3) 35 years of age, 4) patients with prediabetes (should be tested yearly), and 5) people with HIV. On the other hand, in asymptomatic children and adolescents, the screening criteria are: 1) overweight/obesity, 2) maternal history of diabetes/GDM, 3) race/ethnicity, and 4) signs of insulin resistance such as hypertension, dyslipidaemia, polycystic ovary syndrome (PCOS), small for age (American Diabetes Association Professional Practice Committee 2022).

The diagnosis for GDM is done when fasting serum glucose meets or exceed 92 mg/dL, one-hour plasma glucose exceeds 180 mg/dL, and two-hour plasma glucose exceeds 153 mg/dL (American Diabetes Association Professional Practice Committee 2022).

5.6 AETIOLOGY

5.6.1 GENETIC FACTORS

Genetics play a key role in the aetiology and pathogenesis of both T1DM and T2DM. Genetic mutation and associated genetic anomalies are one of the major reasons leading to diabetes. Various genes like the Human Leuckocyte Antigen (HLA) gene and non-HLA genes are associated with autoimmunity and immune-mediated pancreatic β-cell destruction in T1DM (Banday et al. 2020). Heredity also plays a role in acquiring T2DM. The lifetime risk for developing T2DM increases from 40% to 70%, depending on whether either one or both parents have T2DM. Genes influence the risk of developing diabetes by directly influencing insulin secretion, insulin action, and alteration in gene–environment interaction. Sometimes, non-genetic factors influence gene expression. For instance, the gene expression, maternal environment influencing intrauterine life increases the risk of T2DM inheritance. Nevertheless, the development of diabetes is a result of the interplay of genetics, epigenetics, and environment (Ali 2013).

5.6.2 ENVIRONMENTAL FACTORS

5.6.2.1 Viruses

Viral attack during foetal life increases the susceptibility of acquiring T1DM in childhood. Viral infections, especially enterovirus infection, evoke immune response. These viruses can initiate or speed up islet autoimmunity (Coppieters et al. 2012; Paschou et al. 2018). Likewise, hepatitis C virus is associated with T2DM (Toniolo et al. 2019). Molecular mimicry, inflammation, endoplasmic reticulum stress, and suppression of T-cells are the postulated mechanisms by which viral infection leads to the destruction of islet cells (Giwa et al. 2020).

The recent COVID-19 epidemic has increased the incidence of T2DM, probably because of cytokines and Tumour Necrosis Factor-α-induced β-cell dysfunction and insulin resistance at the time of remission of the COVID infection (Rathmann et al. 2022)

5.6.2.2 Gut Microbiota

Dysbiosis is associated with diabetes. Alteration in the composition of gut microbiota can affect intestinal permeability, modulate the immune response, increase weight and insulin resistance, and may lead to T1DM (Paschou et al. 2018), T2DM (Toniolo et al. 2019) and GDM (Plows et al. 2018).

5.6.2.3 Seasonality

Seasonality is also correlated with the incidence of T1DM. Though not very conclusive, it has been seen that incidence rates are higher in cold and temperate seasons, probably due to a higher rate of viral infections during those seasons (Giwa et al. 2020). The effect of seasonality was also seen for GDM. The incidence of GDM increased during the summer and declined during cold months (Chiefari et al. 2017).

5.6.2.4 Diet

Short duration of breastfeeding, early introduction of cow's milk, untimely weaning with cereal, and vitamin D deficiency increase the chances of developing T1DM in children (Paschou et al. 2018; Giwa et al. 2020). A poor diet including food items high in glycaemic load, saturated and trans fats, and low in cereal and polyunsaturated fatty acids (PUFA) has been associated with a higher risk of diabetes in women (Hu et al. 2001).

Dietary patterns consisting of intake of red and processed meat, white rice, potatoes and fries, sugar-sweetened beverages, fruit juices, sweets, desserts, animal products (Ardisson Korat et al. 2014; Adeva-Andany et al. 2019), refined cereals, fructose-rich food items, overcooked food products especially made from refined cereals, aerated soft drinks, energy-dense sugar-and fat-rich meals (Raman 2016) were associated with T2DM. The same holds true for GDM (Plows et al. 2018). Such food items reduce insulin sensitivity.

Besides meal composition, the quality and quantity of food, and the method of cooking the food also plays an important role in predisposition to diabetes. Foods cooked at high temperatures like roasted and fried food, dairy products with sugar form advanced glycation and lipooxidation end products (AGEs and ALEs) contribute to insulin resistance (Raman 2016). Consumption of vegetables, which is considered to have a protective effect, may become a risk factor depending upon the method of cooking. For instance, eating vegetables as steamed or boiled may have a protective effect, whereas the stir-fried form may be deleterious (Xia et al. 2021).

The association of individual nutrients with diabetes was also studied. Vitamin D regulates calcium, which in turn controls the release of insulin, and hence, deficiency of vitamin D becomes a risk factor for the development of diabetes. Deficiency of protein is related to β-cell dysfunction, and therefore it is seen that children who have been subject to protein deficiency during infancy and foetal life are at risk. Similarly, protein-deficient people consuming cyanogenic glycoside-containing foods like cassava, sorghum, yam, and some varieties of beans as a staple diet cannot inactivate the toxin hydrocyanic acid released from cyanogenic glycoside, because of the deficiency of protein. Consequently, the hydrocyanic acid damages the pancreas, leading to diabetes (Raman 2016). Deficiency of certain trace elements, such as boron, cobalt, iodine, iron, magnesium, selenium, and zinc is associated with oxidative stress, which in due course leads to insulin resistance (Dubey et al. 2020).

Nutritional transition has played a major role in increasing the diabetic epidemic. The increase in availability, affordability, and consumption of refined high fat-sugar fast foods and the decline of traditional healthy diets is one of the key drivers in increasing the incidence of diabetes (Ley et al. 2014)

5.6.2.5 Obesity

Obesity is one of the major risk factors in the development of T2DM. Obesity increases the risk of developing T2DM by 5.8 times compared to individuals with normal weight (Diabetologia 2019). The risk of developing diabetes increases with an increase in body mass index (BMI). The risk is found to be 28 times greater in women with a BMI of 30 kg/m^2 and 93 times greater in women with a BMI of 35 kg/m^2 compared to women with normal weight (Barnes 2011).

Obesity adversely influences insulin sensitivity and β-cell function. In obese persons, non-esterified fatty acids are released, which leads to insulin resistance and β-cell dysfunction responsible for the development of T2DM (Al-Goblan et al. 2014).

5.6.2.6 Sedentary Behaviour

Sedentary behaviour is associated with the risk of developing diabetes. Physical inactivity adversely affects the expression, translocation, and function of genes/proteins associated with glucose homeostasis. They suppress insulin signalling leading to insulin resistance. Physical inactivity also lowers β-cell sufficiency, mitochondrial oxidative capacity, and increases oxidative stress and inflammation, resulting in the suppression of insulin sensitivity and, consequently, the development of diabetes (Yaribeygi et al. 2021).

5.6.2.7 Smoking

Current smoking increases the risk of diabetes (Hu et al. 2001). Both active and passive smokers have a higher risk of diabetes (Ardisson Korat et al. 2014). Passive smokers were reported to be at a 22% higher risk than non-smokers who were never exposed to passive smoke. Smoking during pregnancy increases the risk of GDM. In smokers, the risk of diabetes reduces as for non-smokers after ten years of quitting smoking. Nicotine found in cigarettes alters glucose homeostasis, fat distribution, and insulin sensitivity (Maddatu et al. 2017).

5.6.2.8 Alcohol

Alcohol consumption and its risk with the development of diabetes and its complications are quite debatable. Studies have shown that abstinence from alcohol increases the risk of diabetes (Hu et al. 2001) and moderate consumption of alcohol reduces the risk of diabetes (Ardisson Korat et al. 2014) but conversely, it has also been reported that regular consumption of alcohol, even in moderate amounts, interferes with blood sugar control and increases the risk of associated complications like retinopathy and peripheral neuropathy. Alcohol consumption has a deleterious effect on diabetics. Heavy alcohol consumption causes ketoacidosis and dyslipidaemia, and heavy drinking in the fasting state leads to hypoglycaemia and even death (Emanuele et al. 1998). Baliunas et al. (2009) concluded a U-shape relationship between the relative risk for diabetes and alcohol consumption. They reported that in both men and women, moderate consumption of alcohol (two drinks/day) had a protective effect against diabetes compared to abstainers; however, the risk increased with higher levels of consumption (above 50 g/day for women and 60 g/day for men). On the other hand, Knott et al. (2015) concluded that the protective effect of moderate alcohol consumption is confined to women and the non-Asian population.

5.6.2.9 Sleep

Disturbances in sleep are also associated with diabetes. Inappropriate sleep duration and snoring have been seen to increase the risk of developing diabetes (Ardisson Korat et al. 2014). People with sleep disturbances had a 2–3 times higher risk of developing diabetes, probably because increased sympathetic nervous activity is associated with sleep disturbance, which causes glucose intolerance and consequently diabetes (Kawakami et al. 2004). Increased levels of cortisol, inflammation, fatigue, disruption of energy metabolism, loss of β-cell functioning, and apoptosis of β-cells are other postulated mechanisms resulting in insulin resistance because of sleep disturbance (LeBlanc et al. 2018). Therefore, disrupted sleep quality and duration act as risk factors for diabetes.

5.6.2.10 Drugs

Certain medications increase the risk of diabetes. It has been seen that some antidepressant medications, atypical antipsychotic medications, lipid-modifying agents, chemotherapeutic agents, oral contraceptives, and hypertensive medication such as thiazide diuretic and β-blocker increase the risk of diabetes as these drugs can alter the insulin sensitivity as well as secretion and may have a direct cytotoxic effect on pancreatic cells (Ardisson Korat et al. 2014; Fathallah et al. 2015; Repaske 2016).

5.6.2.11 Medical Conditions/Diseased State

The risk of T2DM increases in middle-aged women with PCOS (Gambineri et al. 2012). The rate of T2DM was two times higher in women with PCOS compared to their non-PCOS counterparts (Persson et al. 2021). Hypertension is a potent risk factor for T2DM (Kim et al. 2015). Dyslipidaemia increases the chances of T2DM. It was seen that subjects with higher levels of triglycerides, cholesterol, and lower levels of high-density lipoprotein (HDL) were more prone to develop diabetes compared to their healthy counterparts (Peng et al. 2021). Early menarche at the age of 8–11 years increases the risk of T2DM and GDM, for which one of the probable reasons is association of early puberty with an increase in adiposity (Elks et al. 2013; Xiao et al. 2017). Preterm birth has been associated with both T1DM and T2DM (Li et al. 2014), possibly because of a reduction in the number and functioning of pancreatic β-cells, alteration in immune function and insulin resistance (Crump et al. 2020). Low or high weight during birth is a strong predictor of GDM in later life (Innes et al. 2002).

5.6.2.12 Psychological State

Loneliness and depression are associated with T2DM risk (Hackett et al. 2020). Similar conclusions were drawn by Lindekilde et al. 2021. They found that anxiety, depression, and insomnia increased the incidence of T2DM and attributed the association to biological, cognitive, and behavioural mechanism.

5.6.2.13 Environmental Exposure to Pollutants and Chemicals

Environmental exposure to certain chemicals contributes to diabetes risk. Exposure to inorganic arsenic, organochlorine pesticides, dioxins, and polychlorinated biphenyls has been seen to be positively associated with diabetic risks. These chemicals are postulated to alter gene expression resulting in insulin resistance and are toxic to pancreatic β-cells, thereby impairing the production and secretion of insulin. Exposure to tobacco smoke and air pollution is also associated with diabetes risk. The mixtures of chemicals present in smoke and air pollutants promote inflammation, endothelial dysfunction, and insulin resistance (Starling and Hoppin 2015). Exposure to perfluoroalkyl substances increases the risk of GDM (Yu et al. 2021).

Increased exposure to noise also increases the risk of diabetes, probably by increasing stress level and disturbances in sleep (Shin et al. 2020). Poor housing conditions, unsafe neighbourhoods, and less green space increase the risk of diabetes as they may restrict physical activity and increase stress, which consequently has an impact on insulin sensitivity (Dendup et al. 2018).

5.6.2.14 Ethnicity

Ethnicity/race contributes to the discrepancy in the prevalence rate of diabetes. It was seen that the prevalence of T2DM was higher in Asian and Black ethnic groups compared to white (Pham et al. 2019). The highest prevalence of diabetes was reported in Native Americans (33%). The prevalence rate was higher in non-Hispanic Blacks (12.6%) and Hispanic Americans (11.8%) compared to non-Hispanic whites (7.1%), Asian-Americans (8.4%), and Alaska Natives (5.5%). Race/ethnic disparities were also seen in macro- and microvascular complications of diabetes. Biological factors like insulin sensitivity, obesity, genetic architecture, and non-biological factors were the contributors to increasing the prevalence of diabetes among race/ethnic minorities (Spanakis and Golden 2013).

5.6.2.15 Socioeconomic Factors

Age, gender, low level of education, and financial status are predictors of diabetes. Age is one of the non-modifiable risk factors for T2DM. A significant increase in the prevalence of T2DM was observed with an increase in age, and the ageing effect was most pronounced after 40 years of age (Fazeli et al. 2020). Likewise, advancing maternal age was positively associated with the risk of GDM (Innes et al. 2002). Ageing is associated with an impairment in energy homeostasis, carbohydrate metabolism, β-cell functioning, and a decrease in insulin secretion and insulin sensitivity. Additionally, an increase in visceral fat and the prevalence of cortisol due to ageing promotes insulin resistance leading to diabetes (Mordarska and Godziejewska-Zawada 2017) Gender also influences the prevalence of diabetes. It was seen that females outnumbered males at pubertal age, perhaps due to hormonal influence, and males outnumbered females during adulthood, perhaps due to more endocrine-active visceral fat. No such gender difference was seen in older age, and after age 60, the risk of diabetes was equal for both males and females (Awa et al. 2012). Low educational levels were associated with an increased risk of diabetes, perhaps because higher education creates awareness and motivates individuals to adopt healthy behaviours. Low income levels were associated with an increased risk of diabetes as it may restrict accessibility to healthcare services and promote social deprivation (Hwang and Shon 2014; Duan et al. 2022). Contrarily, it was also seen that higher educational attainment and greater household wealth were associated with an increased risk of diabetes compared to those who had no formal schooling, probably because of unhealthy changes in diet and lifestyle (Seiglie et al. 2020). Thus, it is quite unclear and needs further studies to draw conclusive statements.

5.7 MANAGEMENT OF DIABETES

Apart from treating diabetes and its associated complications, there is an immediate need to make stringent efforts to prevent the global diabetes boom. The modifiable risk factors contributing to diabetes need timely checks to control the upsurge (Figure 5.2).

5.7.1 Pharmacological Management

Since the primary cause of T1DM is insulin deficiency because of the autoimmune destruction of β-cells, insulin therapy is the most recommended therapy. The forms of insulin available are human

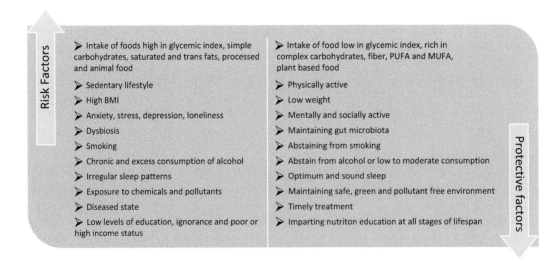

FIGURE 5.2 Modifiable risk factors and protective factors in managing diabetes.

insulin and insulin analogues. Human insulin is 1) short acting, 2) intermediate acting, and 3) pre-mixed. The insulin analogues are 1) rapid acting, 2) long acting, 3) premixed insulin analogues, and 4) co-formulation (ICMR 2018). The administration of insulin varies according to the need of maintaining normoglycaemia. Long-acting insulin analogues or human intermediate-acting are given 1–3 times in the fasting state and between meals. Rapid-acting insulin analogues or short-acting insulin are given before each meal and at a time when correction in blood glucose is required. Premixed insulin is not given in T1DM. Rapid-acting inhaled insulin is available but it has its own constraints, preventing use at a wider level. It is suggested that basal insulin analogues are better in maintaining glycaemic control and reduce the risk of hypoglycaemia and weight gain in T1DM (Janez et al. 2020). Besides insulin therapy, antihyperglycaemic drugs like biguanides, sodium-glucose co-transporter 2 inhibitors, amylin analogue, incretin-based agents are used as adjunctive therapy to provide nephroprotection, vascular endothelium protection, lower insulin resistance, and insulin dose and maintain weight in type 1 diabetics (Otto-Buczkowska and Jainta 2017).

Antihyperglycaemic drugs have been approved for maintaining blood glucose levels and slowing the progression of complications associated with T2DM. Biguanides, sulphonylureas, dipeptidyl peptidase-4 inhibitors, sodium glucose transporter 2 inhibitors, thiazolidinediones, alpha glucosidase inhibitors, non-sulphonylurea insulin secretagogues are some of the approved drugs for treating diabetes. These drugs function through multiple mechanisms such as increasing the sensitivity of the liver and peripheral tissues to insulin, stimulating insulin secretion from β-cells, uptake of glucose by tissues, improving lipid profile, promoting weight loss, inhibiting hepatic glucose output, suppressing the release of glucagon, delaying carbohydrate intake from the gastrointestinal tract, thereby decreasing post-prandial glucose. The combination of these drugs and insulin analogues is also used in T2DM, depending upon the glycaemic target to be achieved (ICMR 2018; Chan and Abrahamson 2003). Sodium glucose transporter 2 inhibitors and glucagon-like peptide analogues are recommended for diabetic patients with cardiovascular disease and/or associated renal disease (Garg et al. 2020).

Despite the antihyperglycaemic effect of drugs, the side effects and high cost associated with them call for other measures to manage diabetes.

5.7.2 PANCREAS AND PANCREATIC ISLET TRANSPLANTATION

Transplantation of the pancreas or islet cells has emerged as the medical strategy to achieve normoglycaemia, especially in individuals with T1DM. Pancreas transplantation is an invasive technique in which the pancreas is transplanted either as simultaneous pancreas–kidney transplant, a pancreas after kidney transplant, or pancreas transplant alone, depending on the severity of the disease and its associated complications. Pancreas transplant is effective in achieving glycaemic control and providing insulin independence. However, the surgical and other risks associated with it do not make it popular among the paediatric population. Thus, islet transplantation can be a promising substitute (Kochar and Jain 2021).

Islet transplantation is a technique in which pancreatic islets isolated from donors are percutaneously infused into the liver via the portal vein of the diabetic subject. This treatment for insulin-dependent diabetes has advantages of reducing the surgical risk, toxicity of immune therapy, and clinical, economic burden of the patient. However, the post-implantation risks call for further research in this arena (Kochar and Jain 2021;Cayabyab et al. 2021)

5.7.3 MEDICAL NUTRITION THERAPY (MNT)

It is a nutrition-based treatment provided by a registered dietitian. It includes nutrition diagnosis, therapeutic nutrition counselling. The therapeutic approach includes nutrition assessment, followed by setting practical goals, nutrition intervention which includes an individualised diet plan and nutrition education, and finally evaluation. The individualised diet is designed as per the individual's age, sex, BMI, cultural practice, and physical activity. Various working groups on diabetes have given recommendations for MNT for managing diabetes. According to the Indian Council of Medical Research, it is recommended that energy intake should be sufficient to maintain the ideal body weight in the case of T2DM, of which 50–60% should come from carbohydrates (preferably complex), 12–15% from protein (preferably vegetable source), and 20–30% from fat (preferably MUFA>PUFA>saturated fat). Excess salt of more than 5 g/day, alcohol, and nutritive sweeteners should be avoided. In case of diabetic complications, further modifications in the diet are recommended (ICMR 2018). The American Dietetic Association has recommended that 15–20% of energy should come from protein and restrict saturated fat to <10% of total daily calories, substitution of nutritive sweeteners with non-nutritive sweeteners, restriction of salt to 2.3 g/day, and intake of alcohol in moderation (Viswanathan et al. 2019). Likewise, in GDM, the aim is to maintain maternal and foetal health, gestational weight gain, normoglycaemia, and absence of ketone bodies. Meals are divided into three small-to-moderate meals and two to four snacks. In case of young diabetics (T1DM), the aim of MNT is to maintain normal growth and development and normoglycaemia, avoid hypoglycaemia and diabetic complications. The main aim is to promote lifelong healthy eating habits according to culture, socioeconomic status, family traditions, food preferences, and psychosocial requirement (Kizilgul et al. 2017). This is important to ensure long-term adherence to healthy living.

MNT helps to achieve normoglycaemia, decrease diabetic complications, and promote adherence. MNT has been found to be more effective in achieving glycaemic control, lowering the diabetic risk score, and bringing positive changes in healthy eating behaviour than the usual care group (Parker et al. 2014). Further, it was seen that in overweight and obese T2DM patients who followed a well-defined structured dietary plan had better outcomes in terms of reduction in body weight, body fat percentage, waist circumference, and glycosylated haemoglobin compared to the group which received only recommendations for developing an individualised eating plan (Mottalib et al. 2018).

In a recent study, it was recommended to make the dietary guidelines more precise according to an individual's genetic makeup, especially in the case of T2DM. The genetically guided MNT provided better glycaemic control as the macro and micronutrients and oral nutrient supplementations could be optimised as per individual's requirements. The adherence to genetically informed dietary recommendations was seen to be higher, which is essential for better health outcomes (Gkouskou et al. 2022).

The practical tips regarding nutrient intake for diabetics include inclusion of fibre-rich low glycaemic index food, PUFA and MUFA fats with the elimination of trans fat and restriction of saturated fats, more plant protein, and lean animal protein.

However, there is no fixed dietary regime for managing diabetes. As discussed above, it varies from individual to individual depending on their phenotype and genetic makeup.

5.7.4 Antioxidants

Antioxidant supplementation may reduce the oxidative stress which plays a crucial role in the pathogenesis of diabetes. It has been seen that supplementation of purified anthocyanin, a natural antioxidant, lowered fasting glucose level, improved dyslipidaemia, increased antioxidant capacity and attenuated insulin resistance in T2 diabetics compared to placebo (Li et al. 2015). Dietary intake of vitamin E and total carotenoids has an inverse relation with the development of T2DM (Jukka et al. 2004). Although certain micronutrients act as antioxidants, their dosage needs to be specified before taking them as supplements by diabetics. However, it is advisable to obtain them from natural food sources as part of the diet rather than supplements.

5.7.5 Herbs and Herbal Drugs

Various herbs and herbal medicines are found to have an anti-diabetic effect. Medicinal plants such as *Allium sativum*, *Aegle marmelos*, *Allium cepa*, *Eugenia jambolana*, *Momordica charantia*, *Ocimum sanctum*, *Phyllanthus amarus*, *Pterocarpus marsupium*, *Tinospora cordifolia*, *Trigonella foenum graecum, Withania somnifera, Urtica, Carthamus tinctorius, Ferula assa-foetida, Bauhinia, Gymnema sylvestre, Swertia, Combretum, Sarcopoterium, Liriope, Caesalpinia bonduc, Coccinia grandis, Syzygium cumini, Mangifera indica, Salvia officinalis, Panax, Cinnamomum verum, Abelmoschus moschatus, Vachellia nilotica, Achyranthes, Fabaceae, Mentha, Asphodelaceae, Andrographis paniculata L., Artemisia herba-alba, Artemisia dracunculus, Azadirachta indica, Caesalpinioideae, Pachira aquatic, Gongronema latifolium, Nigella sativa, Chrysanthemum morifolium, Zingiber zerumbet, Symphytum, Cactaceae, Symplocos, Perilla frutescens, Terminalia chebula*, and *Aloe vera* are observed to be beneficial in diabetes. The whole plant or its parts such as seeds, leaves, bulb, root, aerial parts, stalk, fruit, or gum have the potential to lower blood glucose levels and complications of diabetes (Modak et al. 2007; Moradi et al. 2018). However, it is said that the selection of these herbal drugs depends upon the severity of diabetes or the existing comorbidities. For instance, *Aloe vera* gel and *Nigella sativa* enhance the level of insulin, cinnamon is suitable in the presence of hypertension as a comorbidity, curcumin prevents the progression of the disease, *Urtica* leaves control insulin resistance; cinnamon, China aster, mistletoe fig, and bitter oleander have α-glucosidase-inhibiting action and some resistance, and herbs like *Allium sativum, Panax ginseng, Catharanthus roseus, and Ocimum tenuiflorum*, etc. possess multiple antidiabetic properties (Choudhury et al. 2017). These antidiabetic herbs are also available in the form of polyherbal formulations and have the advantage of being safe, ecofriendly, low cost, and having no adverse effects over synthetic drugs (Verma et al. 2018). The polyherbal formulations were found to be as good as (Majeed et al. 2021) or even statistically better than oral hypoglycaemic agents (metformin) in lowering blood sugar and glycosylated haemoglobin (Suvarna et al. 2021). Despite the advantages and efficacy of being used as an adjuvant or sole treatment, there is still a need for validated testing through well-designed clinical trials on human subjects to earn certification and confidence in the efficacy of the antidiabetic herbal preparation.

5.7.6 NUTRITION EDUCATION

Nutrition and health awareness programmes advocating positive lifestyle changes can bring about a permanent change in the behaviour and attitude of an individual and help in better self-control. Ignorance regarding the pathogenesis of diabetes and the importance of lifestyle changes in prevention may increase the incidence of diabetes and impede the management of the disorder. Many times, therapies for controlling diabetes may fail if the patient is not motivated. Diabetes education has been seen to be effective in improving dietary intake and clinical outcomes. The group receiving education was better at managing glycaemic control and lipid profile (Lim et al. 2009). The inclusion of peer-to-peer support in the nutrition education programme was found to be more effective (Thuita et al. 2020). Many studies have shown the efficacy of education programmes in managing diabetes. and therefore there is a need to conduct such programmes on the regular basis for the vulnerable segment. Young children and adolescents can be targeted at the school and college level so that awareness may be created and the scope for prevention may increase.

5.7.7 PHYSICAL ACTIVITY

Engaging in physical activity lowers the risk of diabetes, slows the progression and can even reverse the process. Physical activity, which includes aerobic exercise, resistance exercise, or flexibility exercise, has been seen to be beneficial in increasing insulin sensitivity, maintaining an optimum lipid profile and body composition. It also helps to reduce the risk of hypoglycaemia and other complications associated with diabetes by improving glycaemic control. Engaging in physical activity is beneficial in all subtypes of diabetes, provided it is personalised in terms of duration, timing, type of exercise depending on the type of diabetes, age of the diabetic, and complications present (Colberg et al. 2016). Adults with T1DM and T2DM should undertake 150 minutes or more of moderate to vigorous intensity activity weekly. Moderate to vigorous intensity resistance training for a minimum of two non-consecutive days per week, stretching exercise (to the point of slight discomfort), and light to moderate intensity of balancing exercise are recommended with a frequency of ≥2–3 times per week. Increasing daily movement is recommended, especially for old, out of condition, and unmotivated diabetics. Long hours of sitting should be interrupted every 30 minutes. Each type of exercise has its own benefit in maintaining glycaemic control and other health benefits. However, it has been recommended that for T2DM adults, a combination of aerobic and resistance training is most suitable for glycaemic control and better health outcomes. In the case of type 1 diabetics, it is important to adjust the recommended exercise regime with meal timings. During fasting conditions, more stable glycaemia is achieved. The carbohydrate intake and insulin dosing need to be modified depending on the intensity and duration of exercise to prevent hypoglycaemia. Regular physical activity is recommended for females prior to or during pregnancy. Females at risk of GDM should engage in 20–30 minutes of moderate-intensity exercise, preferably daily (Colberg 2017). The underlying mechanisms for the preventive and curative effect of exercise therapy in diabetes are: 1) increase in glucose uptake and utilisation, 2) improvement in insulin sensitivity, 3) protection of pancreatic β-cells function, 4) increase in lipid hydrolysis oxidation, 5) alleviation of systemic inflammation, 6) optimisation of BMI, and 7) release of cytokines (Yang et al. 2019).

Yoga is advocated to be favourable in managing diabetes. Yoga practices namely, *suryanamaskar, asanas, shuddhi kriyas, pranayam, bandha, mudras,* and *dhyan* are seen to be beneficial in maintaining glucose levels. These yogic practices stimulate insulin production, rejuvenate and improve the efficiency of pancreatic cells, increase insulin receptor expression and glucose uptake, and decrease insulin resistance (Raveendran et al. 2018). Besides, yogic practices are also helpful in reducing stress and other psychological health issues like depression and anxiety (Shohani et al. 2018) which share a bidirectional relationship with diabetes. Yoga is also seen to be effective in maintaining lipid profile and BMI (Shantakumari et al. 2013).

A 24-week structured yoga programme for type 2 diabetes prevention (Yoga-DP) was developed for high-risk people in India. The programme consists of *Shithilikarana Vyayama* (loosening exercises), *Surya Namaskar* (sun salutation exercises), *Asana* (yogic poses), *Pranayama* (breathing practices), and *Dhyana* (meditation) and relaxation practices (Chattopadhyay et al. 2020). The diabetic yoga protocol intervention of 60 min/day for three months showed a significant reduction in post-prandial glucose, low-density lipoprotein (LDL), and waist circumference (Kaur et al. 2021).

Besides the above mentioned, there are a few other modalities used in the management of diabetes. However, their validation is required before recommendation, and further exploration in the arena is required. These modalities are discussed below.

5.7.8 Massage Therapy

Massage therapy can prove to be useful in diabetes. It has been seen to normalise blood glucose levels, improve diabetic neuropathy and diabetic foot ulcers, and reduce muscle tension, stress, anxiety, and blood pressure (Pandey et al. 2011; Bayat et al. 2019).

5.7.9 Acupuncture

Acupuncture may help in diabetes. It was seen that adjusting internal organs and dredging channel electroacupuncture (AODCEA) treatment for two weeks reduced diabetes-related indicators. The acupuncture technique regulated the structure of intestinal flora, which increases intestinal short-chain fatty acids (SCFAs). SCFAs inhibit the production of pro-inflammatory factors, which consequently improves insulin sensitivity (Zhang et al. 2021). The acupuncture therapy was found to be beneficial when used as an adjuvant to metformin compared to metformin monotherapy. The combination of acupuncture and the drug led to a significant reduction in glucose levels and glycosylated haemoglobin levels (Kazemi et al. 2019).

5.7.10 Aromatherapy

Essential oils are also seen to be effective in lowering blood pressure and circulating glucose levels and improving insulin sensitivity (Talpur et al. 2005). The essential oil obtained from rose-scented geranium had hypoglycaemic and antioxidative effects which were comparable to antidiabetes drugs (Boukhris et al. 2012). Aromatherapy was seen to be effective in reducing sleep disorders, fatigue, anxiety, and neuropathic pain in diabetics thereby improving the quality of life (Cicek and Gencer Sendur 2021).

Various other therapies like chromatherapy and hydrotherapy in combination with other therapies like acupuncture or massage therapy are showing potential in managing diabetes complications and psychological states which have causal outcome relationships with diabetes. Lacunas in human clinical research related to the complementary and alternative therapies discussed above are the barriers to validating their significance. More research is required to specify their duration, dosage, intensity, and prove their efficacy as adjuvant or sole therapies.

Some practical measures for managing and preventing diabetes at government and individual levels are:

- Screening and monitoring of overweight and genetically susceptible individuals on a regular basis for providing timely therapeutic attention
- Designing nutrition education programmes and imparting nutrition education at various levels
- Losing weight is important to prevent diabetes. A 5 kg weight loss resulted in a 50% reduction in the risk of diabetes (Pi-Sunyer 2009)
- Timely diagnosis and treating health conditions like hypertension, dyslipidaemia, PCOS, and mental illnesses. With a decrease in 5 mm Hg of systolic BP, the risk of new-onset diabetes reduces by 11% (Nazarzadeh et al. 2021)
- Safe and green neighbourhood, away from noise and air pollution
- Ensuring prenatal care to avoid adverse birth outcomes and more attention should be paid to children with preterm births
- Regular physical activity is necessary. It was seen that a loss of 7% weight through calorie reduction and 150 min/week of moderate physical exercise prevented diabetes (Barnes 2011). High physical activity reduced the risk of diabetes by 35% in men and 24% in women compared to their counterparts who scored low on physical activity scores (Hjerkind et al. 2017)
- Coping with stress is important. Emotional wellbeing may reduce the chances of developing diabetes as diabetes is also generated by stress. Psychological barriers interfere with the management outcomes, and therefore coping strategies will improve self-care and adherence to treatment. Addressing psychological needs improves glycaemic control and quality of life (Kalra et al. 2018)
- Diet is the cornerstone in preventing diabetes. Low GI food should be given emphasis. A healthy meal consisting of whole cereals, legumes, fruits, vegetables and probiotics, low salt, low sugar, low saturated fats, high fibre, and unsweetened and uncarbonated beverages should be promoted
- Making antidiabetic herbs and spices a part of regular meals
- Abstaining from smoking and alcohol consumption

5.8 CONCLUSION

Diabetes mellitus, characterised by an increase in blood sugar, is increasing at an alarming rate throughout the world. The associated complications are not only taking a toll on life but also increasing the health burden. Therefore, there is a need to manage diabetes and its complications.

Treatment modalities include pharmacological and non-pharmacological approaches. However, some of the modalities are debatable in terms of safety, cost-effectiveness, poor adherence, and poor validation. Preventive measures should be the preferred option over curative measures for counteracting the health challenge. Early identification should be the key to prevention. The subjects who are at risk should be screened on a regular basis and appropriate preventive measures should be taken. Besides, the behavioural factors acting as risk factors should be addressed in a timely manner. Making a change in the dietary pattern, including antidiabetic herbs; leading a physically, mentally, and socially active life; getting proper sleep; abstaining from smoking and alcohol; and imparting nutrition education should be an integral part of the preventive regime. Efforts at the government level and awareness at the individual level can work in synergy in managing the disorder.

REFERENCES

Adeva-Andany, M.M., E. Ranal-Muino, M. Vila-Altesor, C. Fernandez-Fernandez, R. Funcasta-Calderon, and E. Castro-Quintela. 2019. Dietary habits contribute to define the risk of type 2 diabetes in humans. *Clinical Nutrition ESPEN* 34: 8–17. https://doi.org/10.1016/j.clnesp.2019.08.002

Al-Goblan, A.S., M.A. Al-Alfi, and M.Z. Khan. 2014. Mechanism linking diabetes mellitus and obesity. *Diabetes, Metabolic Syndrome and Obesity: Targets and Therapy* 7: 587–591. https://doi.org/10.2147/DMSO.S67400

Ali, O. 2013. Genetics of type 2 diabetes. *World Journal of Diabetes* 4, no. 4: 114–123. https://doi.org/10.4239/wjd.v4.i4.114

American Diabetes Association Professional Practice Committee. 2022. Classification and diagnosis of diabetes: Standards of medical care in diabetes-2022. *Diabetes Care* 45, no. Suppl 1: S17–S38. https://doi.org/10.2337/dc22-S002

Ardisson Korat, A.V., W.C. Willet, and F.B. Hu. 2014. Diet, lifestyle, and genetic risk factors for type 2 diabetes: A review from the Nurses' Health Study, Nurses' Health Study 2, and Health Professionals' Follow-up Study. *Current Nutrition Reports* 3, no. 4: 345–354. https://doi.org/10.1007/s13668-014-0103-5

Awa, W.L., E. Fach, D. Krakow, et al. 2012. Type 2 diabetes from pediatric to geriatric age: Analysis of gender and obesity among 120,183 patients from the German/Austrian DPV database. *European Journal of Endocrinology* 167, no. 2: 245–254. https://doi.org/10.1530/EJE-12-0143

Baliunas, D.O., B.J. Taylor, H. Irving, et al. 2009. Alcohol as a risk factor for type 2 diabetes: A systematic review and meta-analysis. *Diabetes Care* 32, no. 11: 2123–2132. https://doi.org/10.2337/dc09-0227

Banday, M.Z., A.S. Sameer, and S. Nissar. 2020. Pathophysiology of diabetes: An overview. *Avicenna Journal of Medicine* 10, no. 4: 174–188. https://doi.org/10.4103/ajm.ajm_53_20

Barnes, A.S. 2011. The epidemic of obesity and diabetes: Trends and treatments. *Texas Heart Institute Journal* 38, no. 2: 142–144.

Bayat, D., A. Mohammadbeigi, M. Parham, M. Hashemi, K. Mahlooji, and M. Asghari. 2019. The effect of massage on diabetes and its complications: A systematic review. *Cresant Journal of Medical and Biological Sciences* 7: 22–28.

Boukhris, M., M. Bouaziz, I. Feki, H. Jemai, A. El Feki, and S. Sayadi. 2012. Hypoglycemic and antioxidant effects of leaf essential oil of *Pelargonium graveolens* L'Hér. in alloxan induced diabetic rats. *Lipids in Health and Disease* 11: 81. https://doi.org/10.1186/1476-511X-11-81

Cayabyab, F., L.R. Nih, and E. Yoshihara. 2021. Advances in pancreatic islet transplantation sites for the treatment of diabetes. *Frontiers in Endocrinology* 12: 732431. https://doi.org/10.3389/fendo.2021.732431

Chan, J.L., and M.J. Abrahamson. 2003. Pharmacological management of type 2 diabetes mellitus: Rationale for rational use of insulin. *Mayo Clinic Proceedings* 78, no. 4: 459–467. https://doi.org/10.4065/78.4.459

Chattopadhyay, K., P. Mishra, N.K. Manjunath, et al. 2020. Development of a yoga program for type-2 diabetes prevention (YOGA-DP) among high-risk people in India. *Frontiers in Public Health* 8: 548674. https://doi.org/10.3389/fpubh.2020.548674

Chiefari, E., I. Pastore, L. Puccio, et al. 2017. Impact of seasonality on gestational diabetes mellitus. *Endocrine, Metabolic & Immune Disorders Drug Targets* 17, no. 3: 246–252. https://doi.org/10.2174/1871530317666170808155526.

Choudhury, H., M. Pandey, and C.K. Hua. 2017. An update on natural compounds in the remedy of diabetes mellitus: A systematic review. *Journal of Traditional and Complementary Medicine* 8, no. 3: 361–376. https://doi.org/10.1016/j.jtcme.2017.08.012

Cicek, S., and E. Gencer Sendur. 2021. Use of aromatherapy in diabetes management. *International Journal of Traditional and Complementary Medicine Research* 2: 115–120. https://doi.org/10.53811/ijtcmr.959642

Colberg, S.R. 2017. Key points from the updated guidelines on exercise and diabetes. *Frontiers in Endocrinology* 8: 33. https://doi.org/10.3389/fendo.2017.00033

Colberg, S.R., R.J. Sigal, J.E. Yardley, et al. 2016. Physical activity/exercise and diabetes: A position statement of the American Diabetes Association. *Diabetes Care* 39, no. 11: 2065–2079. https://doi.org/10.2337/dc16-1728

Coppieters, K.T., T. Boettler, and M. von Herrath. 2012. Virus infections in type 1 diabetes. *Cold Spring Harbor Perspectives in Medicine* 2, no. 1: a007682. https://doi.org/10.1101/cshperspect.a007682

Crump, C., J. Sundquist, and K. Sundquist. 2020. Preterm birth and risk of type 1 and type 2 diabetes: A national cohort study. *Diabetologia* 63, no. 3: 508–518. https://doi.org/10.1007/s00125-019-05044-z

Dendup, T., X. Feng, S. Clingan, and T. Astell-Burt. 2018. Environmental risk factors for developing Type 2 Diabetes Mellitus: A systematic review. *International Journal of Environmental Research and Public Health* 15, no. 1: 78. https://doi.org/10.3390/ijerph15010078

Diabetologia. 2019. Obesity linked to a nearly 6-fold increased risk of developing type 2 diabetes, with genetics and lifestyle also raising risk. *Science Daily*. www.sciencedaily.com/releases/2019/09/190916081455.htm (accessed 27 June 2022).

Drivsholm, T., N. de Fine Olivarius, A.B. Nielsen, and V. Siersma. 2005. Symptoms, signs and complications in newly diagnosed type 2 diabetic patients, and their relationship to glycaemia, blood pressure and weight. *Diabetologia* 48, no. 2: 210–214. https://doi.org/10.1007/s00125-004-1625-y

Duan, M.-J.F., Y. Zhu, L.H. Dekker, et al. 2022. Effects of education and income on incident type 2 diabetes and cardiovascular diseases: A Dutch prospective study. *Journal of General Internal Medicine*. https://doi.org/10.1007/s11606-022-07548-8

Dubey, P., V. Thakur, and M. Chattopadhyay. 2020. Role of minerals and trace elements in diabetes and insulin resistance. *Nutrients* 12, no. 6: 1864. https://doi.org/10.3390/nu12061864

Elks, C.E., K.K. Ong, R.A. Scott, et al. 2013. Age at menarche and type 2 diabetes risk: The EPIC-InterAct study. *Diabetes Care* 36, no. 11: 3526–3534. https://doi.org/10.2337/dc13-0446

Emanuele, N.V., T.F. Swade, and M.A. Emanuele. 1998. Consequences of alcohol use in diabetics. *Alcohol Health and Research World* 22, no. 3: 211–219.

Fathallah, N., R. Slim, S. Larif, H. Hmouda, and C. Ben Salem. 2015. Drug-induced hyperglycaemia and diabetes. *Drug Safety* 38, no. 12: 1153–1168. https://doi.org/10.1007/s40264-015-0339-z

Fazeli, P.K., H. Lee, and M.L. Steinhauser. 2020. Aging is a powerful risk factor for type 2 diabetes mellitus independent of body mass index. *Gerontology* 66, no. 2: 209–210. https://doi.org/10.1159/000501745

Galicia-Garcia, U., A. Benito-Vicente, S. Jebari, et al. 2020. Pathophysiology of type 2 Diabetes Mellitus. *International Journal of Molecular Sciences* 21, no. 17: 6275. https://doi.org/10.3390/ijms21176275

Gambineri, A., L. Patton, P. Altieri, et al. 2012. Polycystic ovary syndrome is a risk factor for type 2 diabetes: Results from a long-term prospective study. *Diabetes* 61, no. 9: 2369–2374. https://doi.org/10.2337/db11-1360

Garg, S.K., S. Brackett, T. Reinicke, and I.B. Hirsch. 2020. New medication for the treatment of diabetes. *Diabetes Technology & Therapeutics* 22: S-149–S-173. http://doi.org/10.1089/dia.2020.2512

Giwa, A.M., R. Ahmed, Z. Omidian, et al. 2020. Current understandings of the pathogenesis of type 1 diabetes: Genetics to environment. *World Journal Diabetes* 11, no. 1: 13–25.

Gkouskou, K.K., M.G. Grammatikopoulou, E. Lazou, D. Sanoudou, D.G. Goulis, and A.G. Eliopoulos. 2022. Genetically-guided medical nutrition therapy in type 2 diabetes mellitus and pre-diabetes: A series of *n*-of-1 superiority trials. *Frontiers in Nutrition* 9: 772243. https://doi.org/10.3389/fnut.2022.772243

Hackett, R.A., J.L. Hudson, and J. Chilcot. 2020. Loneliness and type 2 diabetes incidence: Findings from the English longitudinal study of ageing. *Diabetologia* 63: 2329–2338. https://doi.org/10.1007/s00125-020-05258-6

Hjerkind, K.V., J.S. Stenehjem, and T.I. Nilsen. 2017. Adiposity, physical activity and risk of diabetes mellitus: Prospective data from the population-based HUNT study, Norway. *BMJ Open* 7, no. 1: e013142. https://doi.org/10.1136/bmjopen-2016-013142

Hu, F.B., J.E. Manson, M.J. Stampfer, et al. 2001. Diet, lifestyle, and the risk of type 2 diabetes mellitus in women. *The New England Journal of Medicine* 345, no. 1: 790–797. https://doi.org/10.1056/NEJMoa010492

Hwang, J., and C. Shon. 2014. Relationship between socioeconomic status and type 2 diabetes: Results from Korea National Health and Nutrition Examination Survey (KNHANES) 2010–2012. *BMJ Open* 4, no. 8: e005710. https://doi.org/10.1136/bmjopen-2014-005710

ICMR. 2018. *ICMR Guidelines for Management of Type 2 Diabetes 2018*. ICMR, New Delhi, 1–66.

Innes, K.E., T.E. Byers, J.A. Marshall, A. Barón, M. Orlean, and R.F. Hamman. 2002. Association of a woman's own birth weight with subsequent risk for gestational diabetes. *JAMA* 287, no. 19: 2534–2541. https://doi.org/10.1001/jama.287.19.2534

Janez, A., C. Guja, A. Mitrakou, et al. 2020. Insulin therapy in adults with type 1 diabetes mellitus: A narrative review. *Diabetes Therapy* 11, no. 2: 387–409. https://doi.org/10.1007/s13300-019-00743-7

Jukka, M., K. Paul, J. Ritva, and R. Antti. 2004. Dietary antioxidant intake and risk of type 2 diabetes. *Diabetes Care* 27, no. 2: 362–366. https://doi.org/10.2337/diacare.27.2.362

Kaku, K. 2010. Pathophysiology of Type 2 diabetes and its treatment policy. *Japan Medical Association Journal* 53, no. 1: 41–46.

Kalra, S., B.N. Jena, and R. Yeravdekar. 2018. Emotional and psychological needs of people with diabetes. *Indian Journal of Endocrinology and Metabolism* 22, no. 5: 696–704. https://doi.org/10.4103/ijem.IJEM_579_17

Kaur, N., V. Majumdar, R. Nagarathna, N. Malik, A. Anand, and H.R. Nagendra. 2021. Diabetic yoga protocol improves glycemic, anthropometric and lipid levels in high risk individuals for diabetes: A randomized controlled trial from Northern India. *Diabetology & Metabolic Syndrome* 13, no. 1: 149. https://doi.org/10.1186/s13098-021-00761-1

Kawakami, N., N. Takatsuka, and H. Shimizu. 2004. Sleep disturbance and onset of type 2 diabetes. *Diabetes Care* 27, no. 1: 282–283. https://doi.org/10.2337/diacare.27.1.282

Kazemi, A.H., W. Wang, Y. Wang, F. Khodaie, and H. Rezaeizadeh. 2019. Therapeutic effects of acupuncture on blood glucose level among patients with type-2 diabetes mellitus: A randomized clinical trial. *Journal of Traditional Chinese Medical Sciences* 6: 101–107. https://doi.org/10.1016/j.jtcms.2019.02.003

Kim, M.J., N.K. Lim, S.J. Choi, and H.Y. Park. 2015. Hypertension is an independent risk factor for type 2 diabetes: The Korean genome and epidemiology study. *Hypertension Research: Official Journal of the Japanese Society of Hypertension* 38, no. 11: 783–789. https://doi.org/10.1038/hr.2015.72

Kizilgul, M., M. Mermer, and B. Ucan. 2017. Medical nutrition therapy for special groups with diabetes mellitus. In *Diabetes Food Plan*, edited by V. Waisundara. IntechOpen, London. https://doi.org/10.5772/intechopen.70815

Knott, C., S. Bell, and A. Britton. 2015. Alcohol consumption and the risk of type 2 diabetes: A systematic review and dose-response meta-analysis of more than 1.9 million individuals from 38 observational studies. *Diabetes Care* 38, no. 9: 1804–1812. https://doi.org/10.2337/dc15-0710

Kochar, I.S., and R. Jain. 2021. Pancreas transplant in type 1 diabetes mellitus: The emerging role of islet cell transplant. *Annals of Pediatric Endocrinology & Metabolism* 26, no. 2: 86–91. https://doi.org/10.6065/apem.2142012.006

LeBlanc, E.S., N.X. Smith, G.A. Nichols, M.J. Allison, and G.N. Clarke. 2018. Insomnia is associated with an increased risk of type 2 diabetes in the clinical setting. *BMJ Open Diabetes Research & Care* 6, no. 1: e000604. https://doi.org/10.1136/bmjdrc-2018-000604

Ley, S.H., O. Hamdy, V. Mohan, and F.B. Hu. 2014. Prevention and management of type 2 diabetes: Dietary components and nutritional strategies. *Lancet* 383, no. 9933: 1999–2007. https://doi.org/10.1016/S0140-6736(14)60613-9

Li, D., Y. Zhang, Y. Liu, R. Sun, and M. Xia. 2015. Purified anthocyanin supplementation reduces dyslipidemia, enhances antioxidant capacity, and prevents insulin resistance in diabetic patients. *The Journal of Nutrition* 145, no. 4: 742–748. https://doi.org/10.3945/jn.114.205674

Li, S., M. Zhang, H. Tian, Z. Liu, X. Yin, and B. Xi. 2014. Preterm birth and risk of type 1 and type 2 diabetes: Systematic review and meta-analysis. *Obesity Reviews: An Official Journal of the International Association for the Study of Obesity* 15, no. 10: 804–811. https://doi.org/10.1111/obr.12214

Lim, H.M., J.E. Park, Y.J. Choi, K.B. Huh, and W.Y. Kim. 2009. Individualized diabetes nutrition education improves compliance with diet prescription. *Nutrition Research and Practice* 3, no. 4: 315–322. https://doi.org/10.4162/nrp.2009.3.4.315

Lin, X., Y. Xu, X. Pan, et al. 2020. Global, regional, and national burden and trend of diabetes in 195 countries and territories: An analysis from 1990 to 2025. *Scientific Reports* 10, no. 1: 14790. https://doi.org/10.1038/s41598-020-71908-9

Lindekilde, N., F. Rutters, J. Erik Henriksen, et al. 2021. Psychiatric disorders as risk factors for type 2 diabetes: An umbrella review of systematic reviews with and without meta-analyses. *Diabetes Research and Clinical Practice* 176: 108855. https://doi.org/10.1016/j.diabres.2021.108855

Maddatu, J., E. Anderson-Baucum, and C. Evans-Molina. 2017. Smoking and the risk of type 2 diabetes. *Translational Research: The Journal of Laboratory and Clinical Medicine* 184: 101–107. https://doi.org /10.1016/j.trsl.2017.02.004

Majeed, M., A. Majeed, K. Nagabhusahnam, L. Mundkur, and S. Paulose. 2021. A randomized, double-blind clinical trial of a herbal formulation (GlycaCare-II) for the management of type 2 diabetes in comparison with metformin. *Diabetology & Metabolic Syndrome* 13, no. 1: 132. https://doi.org/10.1186/s13098 -021-00746-0

Modak, M., P. Dixit, J. Londhe, S. Ghaskadbi, and T.P. Devasagayam. 2007. Indian herbs and herbal drugs used for the treatment of diabetes. *Journal of Clinical Biochemistry and Nutrition* 40, no. 3: 163–173. https://doi.org/10.3164/jcbn.40.163

Moradi, B., S. Abbaszadeh, S. Shahsavari, M. Alizadeh, and F. Beyranvand. 2018. The most useful medicinal herbs to treat diabetes. *Biomedical Research and Therapy* 5, no. 8: 2538–2551. https://doi.org/10.15419 /bmrat.v5i8.463

Mordarska, K., and M. Godziejewska-Zawada. 2017. Diabetes in the elderly. *Przeglad menopauzalny = Menopause Review* 16, no. 2: 38–43. https://doi.org/10.5114/pm.2017.68589

Mottalib, A., V. Salsberg, B.N. Mohd-Yusof, et al. 2018. Effects of nutrition therapy on HbA1c and cardiovascular disease risk factors in overweight and obese patients with type 2 diabetes. *Nutrition Journal* 17: 42. https://doi.org/10.1186/s12937-018-0351-0

Nazarzadeh, M., Z. Bidel, D. Canoy, et al. 2021. Blood pressure lowering and risk of new-onset type 2 diabetes: An individual participant data meta-analysis. *The Lancet* 398: 1803–1810.

Otto-Buczkowska, E., and N. Jainta. 2017. Pharmacological treatment in diabetes mellitus type 1 - insulin and what else?. *International Journal of Endocrinology and Metabolism* 16, no. 1: e13008. https://doi.org /10.5812/ijem.13008

Pandey, A., P. Tripathi, R. Pandey, R. Srivastava, and S. Goswami. 2011. Alternative therapies useful in the management of diabetes: A systematic review. *Journal of Pharmacy & Bioallied Sciences* 3, no. 4: 504–512. https://doi.org/10.4103/0975-7406.90103

Parker, A.R., L. Byham-Gray, R. Denmark, and P.J. Winkle. 2014. The effect of medical nutrition therapy by a registered dietitian nutritionist in patients with prediabetes participating in a randomized controlled clinical research trial. *Journal of the Academy of Nutrition and Dietetics* 114, no. 11: 1739–1748. https:// doi.org/10.1016/j.jand.2014.07.020

Paschou, S.A., N. Papadopoulou-Marketou, G.P. Chrousos, and C. Kanaka-Gantenbein. 2018. On type 1 diabetes mellitus pathogenesis. *Endocrine Connections* 7, no. 1: R38–R46. https://doi.org/10.1530/EC-17 -0347

Peng, J., F. Zhao, X. Yang, X. Pan, J. Xin, M. Wu, and Y.G. Peng. 2021. Association between dyslipidemia and risk of type 2 diabetes mellitus in middle-aged and older Chinese adults: A secondary analysis of a nationwide cohort. *BMJ Open* 11, no. 5: e042821. https://doi.org/10.1136/bmjopen-2020-042821

Persson, S., E. Elenis, S. Turkmen, M.S. Kramer, E.L.Yong, and I.S. Poromaa. 2021. Higher risk of type 2 diabetes in women with hyperandrogenic polycystic ovary syndrome. *Fertility and Sterility* 116, no. 3: 862–871. https://doi.org/10.1016/j.fertnstert.2021.04.018

Pham, T.M., J.R. Carpenter, T.P. Morris, M. Sharma, and I. Petersen. 2019. Ethnic differences in the prevalence of type 2 diabetes diagnoses in the UK: Cross-sectional analysis of the health improvement network primary care database. *Clinical Epidemiology* 11: 1081–1088. https://doi.org/10.2147/CLEP .S227621

Pi-Sunyer, X. 2009. The medical risks of obesity. *Postgraduate Medicine* 121, no. 6: 21–33. https://doi.org/10 .3810/pgm.2009.11.2074

Plows, J.F., J.L. Stanley, P.N. Baker, C.M. Reynolds, and M.H. Vickers. 2018. The pathophysiology of gestational diabetes mellitus. *International Journal of Molecular Sciences* 19, no. 11: 3342. https://doi.org/10 .3390/ijms19113342

Ramachandran, A. 2014. Know the signs and symptoms of diabetes. *Indian Journal of Medical Research* 140, no. 5: 579–581.

Raman, P.G. 2016. Environmental factors in causation of Diabetes Mellitus. In *Environmental Health Risk: Hazardous Factors to Living Species*, edited by M.L. Larramendy and S. Soloneski. IntechOpen, London. https://doi.org/10.5772/62543

Rathmann, W., O. Kuss, and K. Kostev. 2022. Incidence of newly diagnosed diabetes after Covid-19. *Diabetologia* 65: 949–954. https://doi.org/10.1007/s00125-022-05670-0

Raveendran, A.V., A. Deshpandae, and S.R. Joshi. 2018. Therapeutic role of yoga in type 2 diabetes. *Endocrinology and Metabolism* 33, no. 3: 307–317. https://doi.org/10.3803/EnM.2018.33.3.307

Repaske, D.R. 2016. Medication-induced diabetes mellitus. *Pediatric Diabetes* 17, no. 6: 392–397. https://doi.org/10.1111/pedi.12406

Saeedi, P., I. Petersohn, P. Salpea, et al. 2019. Global and regional diabetes prevalence estimates for 2019 and projections for 2030 and 2045: Results from the International Diabetes Federation Diabetes Atlas, 9th edition. *Diabetes Research and Clinical Practice* 157: 107843. https://doi.org/10.1016/j.diabres.2019.107843

Seiglie, J.A., M.E. Marcus, C. Ebert, et al. 2020. Diabetes prevalence and its relationship with education, wealth, and BMI in 29 low- and middle-income countries. *Diabetes Care* 43, no. 4: 767–775. https://doi.org/10.2337/dc19-1782

Shantakumari, N., S. Sequeira, and R. El deeb. 2013. Effects of a yoga intervention on lipid profiles of diabetes patients with dyslipidemia. *Indian Heart Journal* 65, no. 2: 127–131. https://doi.org/10.1016/j.ihj.2013.02.010

Shin, S., L. Bai, T.H. Oiamo, et al. 2020. Association between road traffic noise and incidence of diabetes mellitus and hypertension in Toronto, Canada: A population-based cohort study. *Journal of the American Heart Association* 9, no. 6: e013021. https://doi.org/10.1161/JAHA.119.013021

Shohani, M., G. Badfar, M.P. Nasirkandy, et al. 2018. The effect of yoga on stress, anxiety, and depression in women. *International Journal of Preventive Medicine* 9: 21. https://doi.org/10.4103/ijpvm.IJPVM_242_16

Spanakis, E.K., and S.H. Golden. 2013. Race/ethnic difference in diabetes and diabetic complications. *Current Diabetes Reports* 13, no. 6: 814–823. https://doi.org/10.1007/s11892-013-0421-9

Starling, A.P., and J.A. Hoppin. 2015. Environmental risk factors for Type 2 diabetes: An update. *Diabetes Management* 5, no. 4: 285–299.

Suvarna, R., R.P. Shenoy, B.S. Hadapad, and A.V. Nayak. 2021. Effectiveness of polyherbal formulations for the treatment of type 2 Diabetes mellitus – A systematic review and meta-analysis. *Journal of Ayurveda and Integrative Medicine* 12, no. 1: 213–222. https://doi.org/10.1016/j.jaim.2020.11.002

Talpur, N., B. Echard, C. Ingram, D. Bagchi, and H. Preuss. 2005. Effects of a novel formulation of essential oils on glucose-insulin metabolism in diabetic and hypertensive rats: A pilot study. *Diabetes, Obesity & Metabolism* 7, no. 2: 193–199. https://doi.org/10.1111/j.1463-1326.2004.00386.x

Thuita, A.W., B.N. Kiage, A.N. Onyango, and A.O. Makokha. 2020. Effect of a nutrition education programme on the metabolic syndrome in type 2 diabetes mellitus patients at a level 5 Hospital in Kenya: "A randomized controlled trial". *BMC Nutrition* 6: 30. https://doi.org/10.1186/s40795-020-00355-6

Toniolo, A., G. Cassani, A. Puggioni, et al. 2019. The diabetes pandemic and associated infections: Suggestions for clinical microbiology. *Reviews in Medical Microbiology: A Journal of the Pathological Society of Great Britain and Ireland* 30, no. 1: 1–17. https://doi.org/10.1097/MRM.0000000000000155

Verma, S., M. Gupta, H. Popli, and G. Aggarwal. 2018. Diabetes mellitus treatment using herbal drugs. *International Journal of Phytomedicine* 10, no. 1: 1–10.

Viswanathan, V., D. Krishnan, S. Kalra, et al. 2019. Insights on medical nutrition therapy for type 2 diabetes mellitus: An Indian perspective. *Advances in Therapy* 36: 520–547. https://doi.org/10.1007/s12325-019-0872-8

WHO. 2020. Diabetes. https://www.who.int/news-room/fact-sheets/detail/ the-top-10-causes-of-death (accessed 28 May 2022).

WHO. 2019. *Classification of Diabetes Mellitus 2019: 1–40*. WHO, Geneva.

Xia, M., K. Liu, J. Feng, Z. Zheng, and X. Xie. 2021. Prevalence and risk factors of type 2 diabetes and prediabetes among 53,288 middle-aged and elderly adults in China: A cross-sectional study. *Diabetes, Metabolic Syndrome and Obesity: Targets and Therapy* 14: 1975–1985. https://doi.org/10.2147/DMSO.S305919

Xiao, Y., R. Chen, M. Chen, et al. 2017. Age at menarche and risks of gestational diabetes mellitus: A meta-analysis of prospective studies. *Oncotarget* 9, no. 24: 17133–17140. https://doi.org/10.18632/oncotarget.23658

Yang, D., Y. Yang, Y. Li, and R. Han. 2019. Physical exercise as therapy for type 2 diabetes mellitus: From mechanism to orientation. *Annals of Nutrition & Metabolism* 74, no. 4: 313–321. https://doi.org/10.1159/000500110

Yaribeygi, H., M. Maleki, T. Sathyapalan, T. Jamialahmadi, and A. Sahebkar. 2021. Pathophysiology of physical inactivity-dependent insulin resistance: A theoretical mechanistic review emphasizing clinical evidence. *Journal of Diabetes Research* 2021: 7796727. https://doi.org/10.1155/2021/7796727

Yau, M., N.K. Maclaren, and M.A. Sperling. 2021. Etiology and pathogenesis of diabetes mellitus in children and adolescents. In *Endotext* [Internet], edited by K.R. Feingold, B. Anawalt, and A. Boyce, et al. South Dartmouth (MA): MDText.com, Inc. https://www.ncbi.nlm.nih.gov/books/NBK498653/

Yu, G., M. Jin, Y. Huang, et al. 2021. Environmental exposure to perfluoroalkyl substances in early pregnancy, maternal glucose homeostasis and the risk of gestational diabetes: A prospective cohort study. *Environment International* 156: 106621. https://doi.org/10.1016/j.envint.2021.106621

Zhang, L., X. Chen, H. Wang, et al. 2021. "Adjusting internal organs and dredging channel" electroacupuncture ameliorates insulin resistance in type 2 diabetes mellitus by regulating the intestinal flora and inhibiting inflammation. *Diabetes, Metabolic Syndrome and Obesity* 14: 2595–2607. https://doi.org/10.2147/DMSO.S306861

6 Cardiovascular Diseases

6.1 INTRODUCTION

Sustainable Development Goal (SDG) 3.4 aims to reduce premature mortality from non-communicable diseases through prevention and treatment and promote mental health and wellbeing by one-third. Among various non-communicable diseases, cardiovascular disease (CVD) is a leading cause of death globally. CVD is a group of diseases of the heart, blood vessels, and circulation. In 2019 alone, an estimated 17.9 million people died due to CVDs, which accounted for 32% of global deaths. It is most prevalent in low- and middle-income countries (LMICs). However, between 2000 and 2019, there was a 27% decline in deaths due to CVD globally. The aetiology is multifactorial, which cumulatively can increase the risk of CVDs. Diet, lifestyle (smoking, alcoholism, etc.), and physical activity all contribute to the risk factors of CVDs, which include diabetes, hyperlipidaemia, hypertension, and obesity. It was found in the US that the subjects who had a myocardial infarction (MI) also had at least one uncontrolled risk factor before atherosclerotic cardiovascular disease (ASCVD) events. Therefore, there is a need for controlling or modifying the identified risk factors. This chapter, therefore, focuses on providing a brief insight into the classification, prevalence, mechanism behind how each aetiological factor can increase the risk of CVD, prevention and control through dietary, therapeutic, complementary therapy, and lifestyle management to reduce the risk of CVD.

6.2 TYPES OF CARDIOVASCULAR DISEASES

Cardiovascular disease is associated with atherosclerosis and plaque formation, rheumatic fever, or genetics (Figure 6.1).

6.3 PREVALENCE

The global burden of CVD data was reported in 2015 by integrating data on disease incidence, prevalence, and mortality, and maps for prevalence (Figure 6.2) and death rate (Figure 6.3) were published. Global CVD prevalence was estimated using modelling software and data from health surveys, prospective cohorts, health system administrative data, and registries. The study also estimated CVD mortality from vital registration and verbal autopsy data.

There were 422.7 million estimated cases of CVD globally in 2015. The data from 1990 to 2015 shows many countries did not show any change in CVD prevalence between 1990 and 2015. Including India, some countries showed a significant decline in age-standardised prevalence.

In 1990, the death rate due to CVD was 12.59 million globally, which increased to 17.92 million in 2015. There was a decrease in the death rate from 1990 to 2015 in all high and some middle-income countries; however, there was no significant change in the death rate in sub-Saharan Africa and other countries globally, which includes Oceania and Southeast Asia, Pakistan, Afghanistan, Kyrgyzstan, and Mongolia. According to WHO statistics, 17.9 million people died from CVDs in 2019, which is 32% of all global deaths that occurred in 2019. Among the 32% of CVD deaths, 85% were from heart attacks and strokes, and mostly from LMICs (https://www.who.int.int/new-room/fact-sheet/detail/cardiovascular-disease-(cvds)).

In India, CVDs contributed to 28.1% of deaths in 2016, within India, Kerala, Punjab, and Tamil Nadu are the states with the highest prevalence of CVDs and also have the highest prevalence

DOI: 10.1201/9781003354024-6

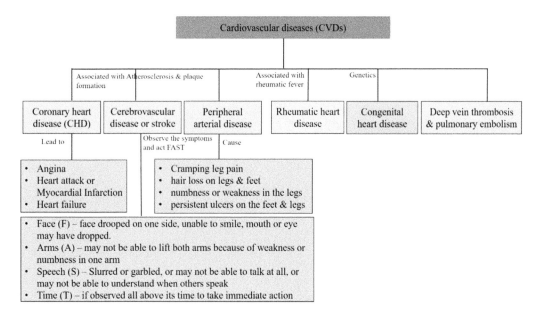

FIGURE 6.1   Types of cardiovascular diseases.

of high cholesterol levels and blood pressure. The prevalence of CVD among diabetic subjects is almost double (21.4%) in India compared to non-diabetic subjects (11%), and the rural population has a lower prevalence of CVD compared to the urban population. Among CVDs, ischemic heart disease and stroke are the major cause of >80% of deaths related to CVDs in India.

6.4   AETIOLOGY

There are multiple factors involved in CVDs, in which some are modifiable and some are non-modifiable. The coexistence of some of the modifiable factors (Figure 6.4) generally aggravates the CVD, however, as these are modifiable, it gives us an opportunity to reduce the risk.

6.4.1   Diabetes, Hyperlipidaemia, and Obesity

Hyperglycaemia and intracellular metabolic changes result in oxidative stress, low-grade inflammation, and endothelial dysfunction, initiating atherosclerotic diabetes. Individuals with diabetes have a two- to four-fold increased risk of developing CVDs due to endothelial dysfunction. Diabetes subjects are most likely will also have other coexisting risk factors such as obesity, hypertension, and dyslipidaemia. All these cumulatively put the subjects at an increased risk of CVD. In the case of type 2 diabetes, subjects with chronic high blood glucose levels cause glycosylation and oxidation, which causes atherosclerotic diabetes. The mechanism behind the association between these risk factors involves low-grade inflammation in the case of obesity. In the case of dyslipidaemia, there are increased free fatty acids and formation of triglycerides, which in turn increases apolipoprotein B secretion and raises the level of very-low-density lipoprotein (VLDL) cholesterol and low-density lipoprotein (LDL-C) cholesterol while decreasing high-density lipoprotein (HDL-C) cholesterol.

6.4.2   Smoking

Tobacco smoking is a modifiable and preventable cause of CVD and other related deaths. Even smoking one cigarette on a daily basis can increase the CVD risk by 40 to 50% compared to those who smoke even more cigarettes. By causing endothelial dysfunction, smoking increases the

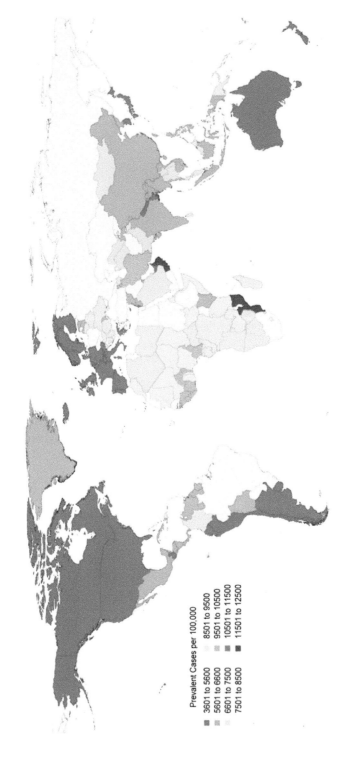

FIGURE 6.2 Age standardised, Prevalence map for CVD in 2015 (Source: Roth et al., 2017).

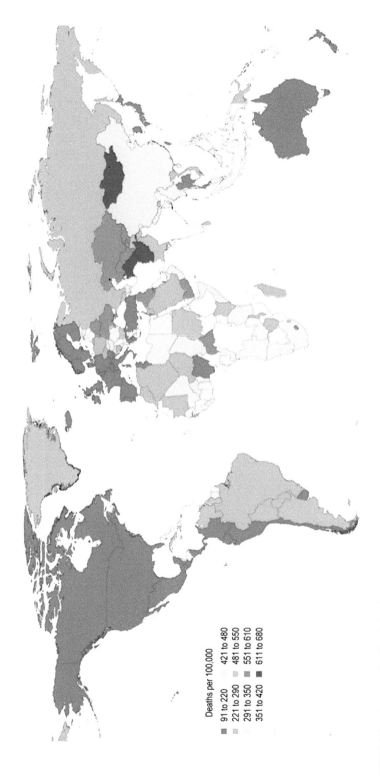

FIGURE 6.3 Age standardised, death rate map for CVD in 2015 (Source: Roth et al., 2017).

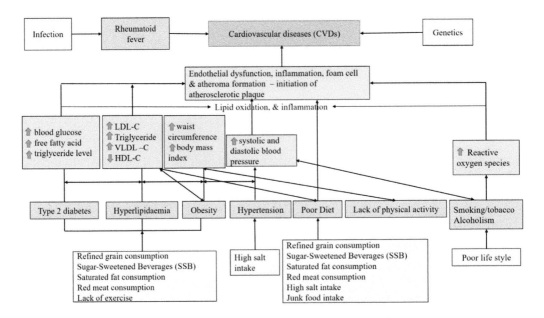

FIGURE 6.4 Etiology of cardiovascular disease.

incidence of atherosclerosis and CVD. The components of cigarette smoke, such as nicotine, tar, and carbon monoxide, impair the vasodilator (nitric oxide, prostacyclin, and hyperpolarising factors) production from the endothelium. Furthermore, reactive oxygen species, specifically superoxide anions formed due to cigarette smoke, reduce the nitric oxide vasodilator and cause oxidation of LDL-C. The cytokines produced due to inflammation at the injured endothelium cause endothelial dysfunction. This process impairs the anticoagulation process and initiates the formation of atherosclerosis. Nicotine from one cigarette can elevate the systolic BP to 20 mmHg and the half-life of nicotine from one cigarette is 2 hours, which means continuous smoking keeps the BP elevated. The study conducted to investigate the effect of smoking on peripheral and central blood pressure shows that the systolic BP was higher in smokers (>5 cigarettes/day), light smokers than in non-smokers (126.7 ± 15.3 (smokers), 127.2 ± 16.5 (light smokers) vs 121.9 ± 13.1 mmHg (non-smokers)). High BP increases the risk of CVD. Smoking also increases the risk of diabetes by damaging β-cell function and causing smoke-related insulin resistance.

6.4.3 Physical Inactivity

It is a well-established risk factor for CVDs. It helps to manage the cardiometabolic risk factors by maintaining fitness, body mass index (BMI), maintaining systolic and diastolic blood pressure, reducing cholesterol level, reducing serum triglyceride level, and improving cardiorespiratory fitness. Initially, the research studies focused on determining the amount of physical activity to calculate the energy expenditure and requirement to perform various tasks, and later it shifted towards maintaining fitness. However, in the past few decades, the focus on physical activity is increasing to maintain health, especially cardiac health. In the year 2000, an important symposium involving 24 experts from six countries gathered at Hockley Valley Resort, near Toronto to evaluate the evidence on the dose–response relationship between physical activity and health outcomes and to write a consensus statement. It states that most of the health benefits are attained at a low to moderate physical activity level, and the greatest health benefits are obtained only when the level of physical activity is high. The review was conducted to collate the science-based evidence on the dose–response relationship between physical activity and CVDs. However, until 2001, no randomised clinical trial was conducted to establish the link based on the dose–response relationship between physical activity and cardiac health. Most of the evidence available was based on prospective observational studies which showed that physical activity is one of the predisposing factors to CVD, especially ischemic heart disease.

6.4.4 Diet

The diet plays a key role in CVD. Most importantly, some of other risk factors of CVD can also can be controlled or reversed by modifying the unhealthy diet to a healthier diet. Increased consumption of processed food that is rich in fat, sugar, and salt increases the risk of diabetes, obesity, hyperlipidaemia, and high blood pressure. Consuming the planetary health diet supports the increased consumption of wholegrains, vegetables, and fruit which could positively impact managing the lipid profile, glucose level, BMI, and waist circumference (Figure 6.5) which could, in turn have positive effect on cardiac health.

6.4.5 Family History

Although there are several modifiable factors such as diet, physical activity, and metabolic disorders that contribute to CVD, there are non-modifiable risk factors specifically, genetics that predict coronary heart disease (CHD) even if the population is habitually active.

6.5 TREATMENT AND PREVENTION

Early detection of CVD risk helps in appropriate management with counselling and medicines along with behavioural change to keep the modifiable risk factors of CVD under control.

6.5.1 Pharmacotherapy

The medications that are commonly used to reduce the effect of one or more risk factors in CVD include aspirin, beta-blockers, angiotensin-converting enzyme inhibitors, and statins.

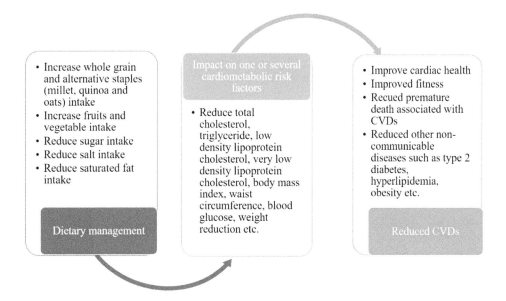

FIGURE 6.5 Dietary management to improve cardiac health.

Primary prevention of cardiovascular events is very important to avoid fatal and disabling events. Initially, aspirin was recommended as a drug for primary prevention of cardiovascular events by some studies. Aspirin's antithrombotic property was observed among tonsillectomy subjects who were bleeding after taking aspirin gums for pain relief, and later it was tested in MI subjects and found to be effective in the prevention of MI. Therefore, it was suggested as a drug for primary prevention of CVDs. However, it is still controversial as bleeding is a major side effect of aspirin, and some studies did not support its effect on CVD. Therefore, it is recommended to use it cautiously and infrequently.

Statin medication is first-line therapy for atherosclerotic CVD in patients with elevated LDL-cholesterol levels, diabetes mellitus, subjects between 45 and 75 years and those at risk of developing ASCVD. In ASCVD, statins lower lipid level by preventing cholesterol synthesis in the liver, and thereby help in reducing the risk and prevent heart disease. Cholesterol-laden plaque deposits in arterial walls affect blood flow and are called atherosclerotic plaque. Statins generally lower blood cholesterol and reduce plaque formation in the arterial wall by inhibiting cholesterol synthesis in the liver. Statins also have a beneficial effect on rheumatoid arthritis-related cardiovascular risk by lowering the lipid level and reducing atherosclerosis, and possibly angioprotective, antioxidative, and anti-inflammatory effects.

For subjects with type 2 diabetes, metformin should be given as a first-line therapy, followed by second-line therapy with a sodium-glucose cotransporter 2 inhibitor or a glucagon-like peptide-1 receptor agonist, if metformin alone is not enough or not tolerated. Metformin improves glycaemic control by inhibiting gluconeogenesis, reducing glucose absorption from the intestine, and increasing glucose uptake by tissues. The inhibition of hepatic gluconeogenesis is achieved by non-competitive inhibition of the mitochondrial glycerophosphate dehydrogenase enzyme, which reduces gluconeogenesis by reducing the conversion of lactate and glycerol to glucose. However, in high-risk patients, additional protection could come from a sodium-glucose cotransporter 2 inhibitor or a glucagon-like peptide-1 receptor agonist.

The systematic review and meta-analysis conducted to determine the effect of angiotensin-converting enzyme inhibitor drugs on cardiovascular events among diabetes mellitus subjects shows that there is a significant reduction (14%) in the risk of major cardiovascular events observed in 14 studies with 34,352 subjects. The study also found therapy with angiotensin-converting enzyme inhibitors reduced MI by 21%, from 11 trials, 22,741 participants with 1944 MI events. Furthermore, this study also reported therapy with angiotensin-converting enzyme inhibitors lowered the risk of heart failure by 19%, from eight trials, 12,651 participants with 782 heart failure occurrences. However, there is no clear evidence of stroke in this study. Another recent systematic review and meta-analysis conducted on COVID-19-infected hypertensive subjects from 26 cohort studies showed that angiotensin-converting enzyme inhibitors/angiotensin receptor blockers treatment significantly reduced mortality in hypertensive subjects.

With the arrival of several drugs for the treatment of hypertension, β-blockers are banished from use as first-line treatment of hypertension. However, β-blockers are still used for the first-line treatment of heart failure, coronary artery disease, and atrial fibrillation as well as hypertension complicated with heart failure, angina pectoris, or prior MI. A recent review conducted on the role of β-blockers in cardiovascular therapy recommended that β-blockers should not be withheld from patients with diabetes mellitus or chronic obstructive pulmonary disease.

The surgical operations that are required to treat CVDs include coronary artery bypass, balloon angioplasty to open the blockage, valve repair and replacement, heart transplantation, and artificial heart operations. Apart from this, the medical devices that are used to treat CVD include pacemakers, prosthetic valves, and patches for closing holes in the heart.

6.5.2 Dietary Management

Among several factors that are associated with CVDs, diet plays a key role and it is a very important factor as it is modifiable, unlike family history associated with CVDs. There is no single food that can be magical in reducing CVD; rather, it requires healthy eating habits with a balanced diet. According to the EAT-Lancet Commission on healthy diets from sustainable food systems, the target for a healthy diet should typically be 300 g vegetables, 250 g dairy, 232 g wholegrains, 200 g fruit, 125 g plant protein, 84 g animal protein, and 50 g tubers and starchy vegetables. However, the global average of consumption of fruit, vegetables, dairy, wholegrains, and plant proteins is under-consumed, whereas animal protein and starchy vegetables are over-consumed. Especially in sub-Saharan Africa, the consumption of tubers and starchy vegetables is extremely high (>300 g vs 50 g target), which is more than six times higher than the target, and all other components of the healthy diet are extremely low, which is alarming. Apart from this, there is a great shift from unprocessed or minimally processed food towards the intake of meat, oil, refined sugar, and refined fat, which is increasing the incidence of diet-related chronic non-communicable diseases, including CVDs.

The recent Lancet Commission report emphasises that among thousands of indigenous foods available globally, only rice, wheat, and maize are consumed on a large scale which reduces the diversity in staple consumption. With the advancement in food-processing technology in recent years, food is processed extensively which not only reduces fibre content and other nutrient content of the food but also increases the risk of developing non-communicable diseases, including CVDs as most of the processed food contains more sugar, salt, and fat. Moreover, rice and wheat are refined to a level that negatively affects the health of the population. Truly speaking, a change in diet in such a way that it is healthier by accommodating more vegetables, fruits, and wholegrains on a daily basis could help manage several non-communicable diseases. Based on eight cohort studies conducted and an analysis to determine diet-dependent percentage reduction in coronary heart disease mortality compared to their regional alternative omnivorous diet, it was reported that Mediterranean, pescatarian, and vegetarian diets reduce the mortality of CHD by around 26%, 20%, and 19% respectively. Although there are various recommendations, below are some general dietary management recommendations supported by scientific studies to prevent, reduce, and/or manage CVDs. It is noteworthy that to maximise the effect of dietary management on CVDs, it is important to follow all the recommendations rather than selecting only one or a few. Dietary guidelines released by the American Heart Association (AHA) to improve cardiovascular health are: 1) Energy intake and expenditure should be adjusted to achieve a healthy body weight. 2) It is important to eat a wide variety of fruit and vegetables. 3) Eat mostly wholegrains rather than refined grains. 4) It is healthy to include protein mostly from plants, fish, and seafood, low-fat and fat-free dairy products, and to choose lean cuts of meat or poultry and avoid processed forms. 5) Instead of tropical oils and partially hydrogenated oils, it is better to use liquid plant oil. 6) Instead of ultra-processed foods, consume minimally processed foods. 7) Minimise intake of beverages and food with added sugar. 8) Prepare food with little or no salt. 9) Limit alcohol intake. Detailed evidence for some of these guidelines is discussed below.

6.5.2.1   Reduce Saturated Fat Intake

It is always healthier to include food that is rich in unsaturated fatty acids, such as fish, nuts, and seeds. Avoid red meat, lard, cream, and fatty junk foods such as cookies, cakes, pizza, and burgers with excessive cheese, cream, and butter.

Dietary saturated fat and abdominal obesity increase LDL-cholesterol levels and are two independent CVD markers. Prospective observational studies were used to conduct the meta-analysis to find the relationship between saturated fat, unsaturated fat, and trans-fat consumption and the risk of mortality from CVD. The result from 12 prospective studies shows a non-linear association between trans-fat consumption and the relative risk of mortality from CVD. For every 5% increase in saturated fat intake, there was a 3% higher risk of mortality from CVD, and the association was non-linear after 12% saturated fat intake. On the other hand, results from 11 prospective cohort studies show 5% energy increment in polyunsaturated fat intake was associated with a 5% reduction in the relative risk of mortality from CVD. Six prospective cohort studies show that a 1% energy increment in trans-fat was associated with a 6% increase in the relative risk of mortality from CVD.

The dose–response meta-analysis using 15 prospective studies found lower cheese consumption was significantly associated with a 10–14% lower risk of CVD. The largest risk reduction was observed at a cheese consumption of 40 g/day, and relative risk of CVD with cheese consumption of 50 g/day was 0.92.

6.5.2.2 Reduce Red Meat Intake

Some studies found a negative association between red meat consumption and CVD, and there is a huge debate on red meat consumption and its association with CVD. Despite this, there are still some studies that find an association between red meat consumption and CVD which cannot be ignored. Red meat is high in both protein and saturated fatty acids. High saturated fat consumption is considered a high-risk factor for metabolic disorders and CVD. The data of 180,642 CVD-free individuals from the UK biobank was used to determine the relative risk of consuming red meat and CVD incidence and mortality. The result shows that the highest intake of red meat consumption increased the risk of CVD mortality by 20%, CHD mortality by 53%, and stroke mortality by 101%, and it was identified as an independent risk factor as the effect was not modified by a change in lifestyle and genetic factors. However, replacing red meat with chicken or cereal reduced the risk of CVD and CHD mortality. A case-cohort study shows a positive association between red meat consumption and increased MI but not with CVD, and identified high ferritin as a marker. The prospective cohort of Sweden with 74,645 men and women evaluated red meat and fruit and vegetable consumption through a self-administered questionnaire, and risk was estimated for CVD, cancer (discussed in another chapter), and all-cause mortality. The result shows that individuals who consume more meat had a 29% and 21% risk of mortality from CVD and all causes, respectively. There was no interaction between red meat consumption and fruit and vegetable consumption, and the risk did not change with high consumption of fruit and vegetables. Hence, fruit and vegetable consumption may not counterbalance the effect of red meat consumption as it could be an independent risk factor.

6.5.2.3 Reduce Salt Intake

There are several pieces of evidence available that show wholegrain consumption is associated with a lower risk of CVDs. However, the global average of wholegrain consumption is less than 50 g compared to the global health diet target of 232 g. On the other hand, refined grain intake is increasing. Wholegrains are rich in fibre, major nutrients, and phytochemicals that are very important to maintain cardiac health.

The Framingham Heart Study (FHS) is the first large and long-term longitudinally followed community-based observational prospective cohort study, initiated in 1948, to investigate the risk factors of CVDs by recruiting 5209 participants. In 1971, a 5124-strong second-generation group was recruited, who were the offspring of the original cohort and they were followed approximately every four years. During the fifth examination cycle in 1991–1995, a dietary assessment was conducted and was considered as baseline, and thereafter data (dietary and cardiometabolic) from the sixth to ninth examinations were used to study the changes in cardiometabolic risk factors such as waist circumference (WC), systolic blood pressure (SBP), diastolic blood pressure, fasting plasma HDL cholesterol, plasma triglyceride, and serum glucose levels of the participants who consumed wholegrain (g/day) or refined grain (g/day). The result shows the greater the wholegrain intake (>48 g/d) the lower the increase in waist circumference (1.4 ± 0.2cm), whereas the lower the wholegrain intake (<8 g/d), the greater the increase in waist circumference (3.0 ±0.1 cm). A similar trend was observed with other cardiometabolic risk factors. Fasting glucose concentration was high among high wholegrain-consuming participants (0.7 ± 0.4 compared with 2.6 ± 0.2 mg/dl) than lower wholegrain-consuming participants. Higher wholegrain consumption also increased HDL cholesterol more compared to low intake of wholegrain (2.4 ± 0.3 vs 2.0 ± 0.2 mg/dL). A greater decrease in triglyceride concentration was observed in high wholegrain-consuming participants compared to a lower decrease in triglyceride concentration in low wholegrain-consuming participants

(−7 ± 1.5 vs −3.8 ± 0.9 mg/dL). On the other hand, greater refined grain intake was associated with a greater increase in waist circumference compared to a lower increase in the low refined grain-consuming category (2.7 ± 0.2 vs 1.8 ± 0.1 cm). High intake of refined grain was also associated with a lower decrease in triglyceride concentration compared to low refined grain intake (−0.3 ± 1.3 vs −7.0 ± 0.7 mg/dL). The study concluded that replacing refined grain with wholegrain among middle-aged to old-aged adults will be effective in attenuating abdominal obesity, dyslipidaemia, and hyperglycaemia and thereby reducing the risk of cardiometabolic diseases.

General dietary guidelines always recommended wholegrain consumption rather than how much wholegrain consumption would be effective in reducing the risk. The dose–response relationship between consumption of wholegrain and its association with CHD and CVD was studied by a systematic review and meta-analysis using the seven cohort studies. The results show that wholegrain consumption reduces the risk of CHD. It was a dose–response meta-analysis, which reported that 90 g, equal to three servings of wholegrain intake per day, reduced the risk of CHD by 19%. However, this slightly steeper reduction was observed only up to three servings a day. Increasing the number of servings didn't end up in a further steep reduction of risk although a further slight reduction of risk was observed up to 210 g per day. Ten cohort studies were used to conduct a meta-analysis to investigate the association of wholegrain consumption with the risk of CVD. The results show there was a non-linear association between wholegrain consumption and the risk of CVD with a stronger reduction in the risk of CVD and intake of wholegrain up to 50 g per day. There was a 22% reduction in the risk of CVD per 90 g consumption per day. The higher intake did not show further reduction although a slight reduction of risk was observed up to an intake of 200 g per day.

A recent systematic review and meta-analysis conducted shows consumption of millet (traditional grain of Asia and Africa) significantly reduces total cholesterol (TC), triglycerides (TG), low-density lipoprotein cholesterol (LDL-C), and very-low-density lipoprotein cholesterol (VLDL-C) compared to other staples. On the other hand, it increases high-density lipoprotein cholesterol (HDL-C) compared to other staples, including refined rice and wheat. It also shows that millet consumption reduces overweight and obesity. Therefore, millets help in managing lipid profile and thereby have the potential to reduce the risk of CVD.

Processed quinoa seeds (although the study did not clearly describe what 'processed' means here and how much it is processed) effect on blood cholesterol levels in 27 CHD and hyperlipidaemia subjects show that consumption of processed quinoa for 120 to 200 days significantly reduced TC, TG, and LDL-C levels.

A systematic review and meta-analysis conducted to study the effect of oat supplementation on CVD risk markers shows that there was an association between oat supplementation and higher reduction of TC, LDL-C, glucose, BMI, body weight, and waist circumference compared to the control group with a regular diet.

The examples above clearly show that wholegrain consumption and alternative grains such as millet, quinoa, and oats could be beneficial in reducing the risk factors of CVDs.

6.5.2.4 Reduce Sugar Intake

The review conducted previously to find the association between sugar intake and CVD risk reported that sugar intake within the normal range of human diet does not have any association with risk for CVDs, except for diets with more than 20% Kcal from simple sugar. This has been shown to increase triglyceride levels, which is one of the CVD risk factors. It is recommended that randomized controlled trials (RCTs) be conducted to prove this association.

The randomised controlled trial conducted to change the dietary behaviour of children and improve maternal weight showed that by reducing the consumption of sugar-sweetened beverage (SSB) juice, overweight or obese mothers (n = 27) lost 2.4 kg weight in six months compared to the mothers in the control group (n = 24) who gained weight of 0.9 kg. This shows SSBs can help in reducing obesity, which is a known risk factor of CVDs.

Sugar levels must be restricted to <10% of energy intake in the interest of reducing CVD risk factors. The prospective study conducted on individuals who consume <10%, 10 to 24.9% and >25% added sugar showed that the hazard ratio of CVD mortality was 1.3 and 2.75 in individuals who consumed added sugar from 10% to 24.9%, and more than 25% respectively compared to reference hazard ratio of CVD which was 1 in individuals who consumed added sugar less than 10%. This clearly shows that an increase in added sugar consumption increases the risk of CVD mortality. A recent randomised controlled cross-over study conducted by testing high sugar and low sugar diet on anthropometry, lipid profile, and glucose level for 12 weeks in 25 adults showed a high sugar diet increased the body weight by 0.7 ± 0.3 kg, waist circumference by 1.4 ± 1.0 cm, fat mass by 0.5 ± 0.3 kg, and plasma triglyceride by 0.26 ± 0.07 mmol/L. The study also revealed that the low sugar diet decreased the body weight by 2.1 ± 0.5 kg, waist circumference by 2.0 ± 0.8 cm, fat mass by 1.4 ± 0.3 kg, and plasma triglyceride by 0.35 ± 0.16 mmol/L. Another recent prospective study conducted among 106,178 women teachers in California to determine the association of SSB intake and CVD risk. There were 8848 CVD incidents over 20 years and the result of consumption of SSBs and risk showed daily consumption of at least one serving of SSB was associated with a higher risk of CVD (1.19), revascularisation (1.26), and stroke (1.21) in Californian women teachers after adjusting for all other potential confounders of CVD. The risk was even higher (1.42) with ≥ one serving of fruit drink/day and caloric soft drink consumption on a daily basis had a higher risk (1.23) of the first CVD event compared to those who never consume or rarely consume. More than 1.5 cups of SSB were associated with MI risk. Although the direct link between sugar intake and CVDs is not proven through a randomised controlled trial, considering that excessive sugar intake can lead to obesity and elevated triglyceride levels, it is safe to minimise the added sugar and excessive dietary sugar. The guidelines on primary prevention of CVD by the AHA also emphasised reducing sweetened beverages in their ten take-home messages.

6.5.2.5 Increase Fruit and Vegetable Intake

Some studies have established the link between fruit and vegetable consumption and the risk of CVDs. In the prospective Kuopio ischemic heart disease risk factor (KIHD) study conducted in Finland, it was found that the higher the fruit and vegetable intake, the lower the risk of mortality related to all causes, CVD, and non-CVD. With increased consumption of fruit and vegetables, there was a decrease in serum total cholesterol and LDL-cholesterol ($p < 0.0001$). There was a slight decrease in body mass index ($p = 0.010$), systolic blood pressure ($p = 0.037$), and diastolic blood pressure ($p = 0.055$). With increased fruit and vegetable intake, there was a significant ($p < 0.0001$) decrease in saturated fat intake and significant ($p < 0.0001$) increase in fibre, vitamin E, folate, vitamin C, and β-carotene intake. The effect was attenuated when the nutritional factors were adjusted, and therefore the effect was attributed to increased fibre, vitamin E, folate, vitamin C, and β-carotene intake. A large-scale population-based cohort study conducted in China showed that the higher the fruit and vegetable consumption the lower the risk of CVDs. Some 100,728 participants were followed for a maximum of 16.5 years with 3677 CVD events. A food frequency questionnaire was used to record the intake of fruit and vegetables. The multivariate hazard risk analysis showed that for every 200 g increase in total fruit and vegetable intake, there was 5% reduction in the risk of developing CVDs. The relationship was non-linear, and the beneficial effect was found for consumption ranging from 0 to 600 g per day, with no additional benefit found beyond consumption of 600 g per day. Only the highest intake of fruit and vegetables was associated with a reduction in all-cause mortality. Data from a study conducted among 65,226 adults above 35 years in the Health Survey for England were analysed to estimate the level of hazard risk and the consumption of fruit and vegetables and their association with CVD and all-cause mortality. Vegetable consumption was shown to have a greater association with the reduction in CVD-related mortality and all-cause mortality, and the study recommends the consumption of more than seven portions of fruit and vegetables. Fruit consumption was not associated with mortality from CVD, especially consuming canned or frozen fruit, which was associated with an increased risk of all-cause mortality.

Highlights

➤ Reduce salt intake to <5 g/day.

➤ Consume whole grain at least 90 g/day.

➤ Reduce sugar level to < 10% of total energy intake to reduce CVDs.

➤ Increase fruits and vegetable consumption (>400g).

FIGURE 6.6 Highlights of dietary management to reduce the risk of CVD.

Dose–response meta-analysis conducted on 17 prospective cohort studies shows a non-linear relationship in a reduction in the relative risk of CHD by 30% with the consumption of 550 to 600 g/day and by a 21% reduction in the relative risk with the consumption of 750 to 800 g per day. Some 64 studies from 58 articles included on CVD show a non-linear association between consumption of fruit and vegetables and CVDs. There were a 28%, 27%, and 28% reduction in the relative risk of CVD with fruit and vegetables, fruit alone, and vegetables alone consumption. The specific fruit and vegetables that show the associated reduction in the relative risk of CVD include apples/pears, citrus fruits, green leafy vegetables/salads, and cruciferous vegetables. The above studies support the fruit and vegetable intake to reduce the risk of CVDs; however, studies on the effect of different types of vegetables and fruit along with large-scale randomised controlled trials are important to generate strong evidence.

Highlights of dietary management to reduce the risk of CVD are provided in Figure 6.6.

6.5.3 COMPLEMENTARY AND ALTERNATIVE THERAPY

Exercise is the most cost-effective and beneficial alternative treatment for CVD compared to other pharmacological and procedural treatments as it exerts multiple benefits such as reducing athero-sclerosis and also helps to manage obesity and diabetes. Generally, it is well known that there is an inverse relationship between physical activity and CHD. However, the amount and intensity of physical activities required dose–response evidence. The Harvard alumni health study was initiated in 1962 with 17,835 men. Some 87% (12,516) of these men were later followed in 1977 and continued until 1993 to examine the association between CHD and the quantity and intensity of physical activity. During this period, 2135 men had CHD incidents, which includes MI, angina pectoris, revascularisation, and coronary death. The results show the relative risk of CHD was significantly (<0.001) less in the men who were expending ≥4200 kJ per week in total physical activity and vigorous activities compared to those who were expending <4200 kJ per week. The most significant cardiovascular risk reduction was observed with increasingly active trajectories of physical activity for more than 20 minutes per day, and it was more marked at 70 years. The physical activity trajectories such as low/decreasing, light/stable, and moderate/increasing were examined for their association with all-cause and CVD mortality by following 3231 men for 20 years. The results show that light/stable group of men had a lower risk (HR; 0.76) of all-cause mortality compared to the low/decreasing (HR; 0.83) trajectory group, and a similar association was observed for CVD mortality.

The systematic review and meta-analysis conducted to determine the effect of yoga on cardiovascular risk parameters shows that yoga reduces diastolic blood pressure. Two studies used in the systematic review and meta-analysis show there was a significant reduction in cholesterol levels with yoga. Five studies in the meta-analysis show there is a significant effect of yoga on increased HDL cholesterol levels and reduced triglyceride levels. However, there is no significant change in LDL-cholesterol levels. The studies used in the above-mentioned systematic review and meta-analysis were short-term studies that require further evaluation with long-term, well-planned randomised controlled trials. The study was conducted to determine the long-term effect of *Sudarshan kriya* on cardiorespiratory physiology among 25 individuals. It showed that there was a significant reduction in cardiovascular parameters such as pulse rate, systolic, and diastolic blood pressure in the individuals practising *Sudarshan kriya* (only) for more than one year compared to the control group who never practised any form of physical exercise or *Sudarshan kriya* regularly. Based on the impact, the study recommended *Sudarshan kriya* as adjuvant therapy for managing hypertension and diabetes. However, the study also suggested conducting a prospective study to see the effect on the same individual before and after *Sudarshan kriya*. Another study conducted on the same *Sudarshan kriya* yoga techniques showed that practising *Sudarshan kriya* had a significant improvement in lipid profile. Thirty subjects who were doing *Sudarshan kriya* showed a significant reduction in arterial blood pressure (from 93.5 ± 4.46 to 90.3 ± 4.33 mmHg), serum cholesterol level (188 ± 24.4 to 180 ± 20.5 mg/dL), and blood sugar level (86.3 ± 6.29 to 82.8 ± 6.03 mg/dL) in a 90-day period. Apart from these studies, there is also a study on the effect of *Sudarshan kriya* on alcoholism that showed there was a significant reduction in the consumption of alcohol after practising *Sudarshan kriya* in six-month period however, its benefit on blood parameters was not studied. Considering the significant reduction in alcohol intake, it leads to the assumption that there could be a change in blood lipid and glucose levels, which needs to be further explored by well-designed interventions.

From the above discussion, it is clear that an active lifestyle helps to maintain cardiac health, and in cardiac treatment it plays a supportive role as a complementary method to improve cardiac health.

6.5.4 LIFESTYLE MODIFICATION

Lifestyle modification is always considered an important aspect of reducing non-communicable diseases. There are some factors that jointly or independently affect cardiac health which need to be modified immediately to reduce the recurrence and secondary prevention of the disease. Increasing physical activity, consuming only a healthy diet, stopping smoking and avoiding second-hand smoke, and addressing the stress which can elevate CVD risk, can jointly with medication or independently help in reducing the risk of CVDs. Along with prescribed medication, it is important that one should also adopt a healthy lifestyle to improve cardiac health. An analytical review conducted recently demonstrated that regular physical activity, sound nutrition, weight management, and not smoking cigarettes had a significant effect on reducing the risk of CVD. By practising a better lifestyle proved to reduce the risk of CVD >80% and diabetes >90%.

6.6 CONCLUSION

There is no one single solution for reducing the risk of CVDs. It should be approached in a collective way considering all modifiable factors such as diet, lifestyle, and complementary methods with or without drugs. Early identification is key in reducing premature mortality. Although the studies on each factor are controversial with some studies supporting and some not supporting it, it is really important to follow work on modifiable factors to keep them under control. With this, more randomised controlled studies are required to strengthen the evidence on each factor and their impact on CVDs.

BIBLIOGRAPHY

Agio, D., E. Papachristou, O. Papacosta, L.T. Lennon, S. Ash, P. Whincup, S.G. Wannamethee, and B.J. Jefferis. 2020. Trajectories of physical activity from midlife to old age and associations with subsequent cardiovascular disease and all-cause mortality. *Journal of Epidemiology Community Health* 74, no. 2: 130–136. https://doi.org/10.1136/jech-2019-212706

Anitha, S., T.W. Tsusaka, R. Botha, J. Kane-Potaka, D.I. Givens, A. Rajendran, and R.K. Bhandari. 2022. Are millets more effective in managing hyperlipidaemia and obesity than major cereal staples? A systematic review and meta-analysis. *Sustainability* 14, no. 11: 6659. https://doi.org/10.3390/su14116659

Arnett, D., R.S. Blumenthal, M. Albert, A. Buroker, Z. Goldberger, E. Hahn, C. Himmelfarb, et al. 2019. 2019 ACC/AHA guideline on the primary prevention of cardiovascular disease. *Journal of the American College of Cardiology* 74, no. 10: e177–e232. https://doi.org/10.1016/j.jacc.2019.03.010

Aryati, A., C. Isherwood, M. Umpleby, and B. Griffin. 2020. Effects of high and low sugar diets on cardiovascular disease risk factors. *Journal of Nutritional Science and Vitaminology* 66: S18–S24. https://doi.org/10.3177/jnsv.66.s18

Aune, D., E. Giovannucci, P. Boffetta, L.T. Fadnes, N. Keum, T. Norat, D.C. Greenwood, E. Riboli, L.J. Vatten, and S. Tonstad. 2017. Fruit and vegetable intake and the risk of cardiovascular disease, total cancer and all-cause mortality—A systematic review and dose-response meta-analysis of prospective studies. *International Journal of Epidemiology* 46, no. 3: 1029–1056. https://doi.org/10.1093/ije/dyw319

Aune, D., N. Keum, E. Giovannucci, L.T. Fadnes, P. Boffetta, D.C. Greenwood, S. Tonstad, L.J. Vatten, E. Riboli, and T. Norat. 2016. Whole grain consumption and risk of cardiovascular disease, cancer, and all cause and cause specific mortality: Systematic review and dose-response meta-analysis of prospective studies. *BMJ*: i2716. https://doi.org/10.1136/bmj.i2716

Baker, C., C. Retzik-Stahr, V. Singh, R. Plomondon, V. Anderson, and N. Rasouli. 2021. Should metformin remain the first-line therapy for treatment of type 2 diabetes? *Therapeutic Advances in Endocrinology and Metabolism* 12. https://doi.org/10.1177/2042018820980225

Barbiellini Amidei, C., C. Trevisan, M. Dotto, E. Ferroni, M. Noale, S. Maggi, M.C. Corti, G. Baggio, U. Fedeli, and G. Sergi. 2022. Association of physical activity trajectories with major cardiovascular diseases in elderly people. *Heart* 108, no. 5: 360–366. https://doi.org/10.1136/heartjnl-2021-320013

Bellavia, A., S. Frej, W. Stilling, and W. Alicja. 2016. High red meat intake and all-cause cardiovascular and cancer mortality: Is the risk modified by fruit and vegetable intake? *The American Journal of Clinical Nutrition* 104, no. 4: 1137–1143. https://doi.org/10.3945/ajcn.116.135335

Bouchard, C. 2001. Physical activity and health: Introduction to the dose-response symposium. *Medicine and Science in Sports and Exercise* 33, no. 6: S347–50. PMID: 11427758. https://doi.org/10.1097/00005768-200106001-00002

Bowles, D.K., and M.H. Laughlin. 2011. Mechanism of beneficial effects of physical activity on atherosclerosis and coronary heart disease. *Journal of Applied Physiology* 111, no. 1: 308–310. https://doi.org/10.1152/japplphysiol.00634.2011

Brinks, J., A. Fowler, B.A. Franklin, and J. Dulai. 2017. Lifestyle modification in secondary prevention: Beyond pharmacotherapy. *American Journal of Lifestyle Medicine* 11, no. 2: 137–152. https://doi.org/10.1177/1559827616651402

Caleigh, M.S., F.J. Paul, H.L. Alice, T.R. Gail, M. Jiantao, S. Edward, and M.M. Nicola. 2021. Whole- and refined-grain consumption and longitudinal changes in cardiometabolic risk factors in the Framingham offspring cohort. *The Journal of Nutrition* 151, no. 9: 2790–2799. https://doi.org/10.1093/jn/nxab177

Cappuccio, F.P. 2013. Cardiovascular and other effects of salt consumption. *Kidney International Supplements* 3: 312–315.

Carnethon, M.R. 2009. Physical activity and cardiovascular disease: How much is enough? *American Journal of Lifestyle Medicine* 3, no. 1: 44S–49S. https://doi.org/10.1177/1559827609332737

Chacon, D., and B. Fiani. 2020. A review of mechanisms on the beneficial effect of exercise on atherosclerosis. *Cureus* 12, no. 11: e11641. https://doi.org/10.7759/cureus.11641

Chen, G.C., Y. Wang, X. Tong, I.M.Y. Szeto, G. Smit, Z.N. Li, and L.Q. Qin. 2017. Cheese consumption and risk of cardiovascular disease: A meta-analysis of prospective studies. *Europian Journal of Nutrition* 56, no. 8: 2565–2575. https://doi.org/10.1007/s00394-016-1292-z

Cheng, J., W. Zhang, X. Zhang, F. Han, X. Li, X. He, Q. Li, and J. Chen. 2014. Effect of angiotensin-converting enzyme inhibitors and angiotensin II receptor blockers on all-cause mortality, cardiovascular deaths, and cardiovascular events in patients with diabetes mellitus: A meta-analysis. *JAMA Internal Medicine* 174, no. 5: 773–785. https://doi.org/10.1001/jamainternmed.2014.348

Daniel, A., P. Quintana, S. Disorn, W. Clemens, E.G. Mirja, S. Ruth, J. Theron, K. Verena, J. Paula, K. Rudolf, and K. Tilman Kühn. 2018. Red meat consumption and risk of cardiovascular diseases—Is increased iron load a possible link? *The American Journal of Clinical Nutrition* 107, no. 1: 113–119. https://doi.org/10.1093/ajcn/nqx014

Dézsi, C.A., and V. Szentes. 2017. The real role of β-blockers in daily cardiovascular therapy. *American Journal of Cardiovascular Drugs* 17: 361–373. https://doi.org/10.1007/s40256-017-0221-8

Gallucci, G., A. Tartarone, R. Lerose, A.V. Lalinga, and A.M. Capobianco. 2020. Cardiovascular risk of smoking and benefits of smoking cessation. *Journal of Thoracic Disease* 12, no. 7: 3866–3876. https://doi.org/10.21037/jtd.2020.02.47

Gleissner, C.A., E. Galkina, J.L. Nadler, and K. Ley. 2007. Mechanisms by which diabetes increases cardio-vascular disease. *Drug Discovery Today: Disease Mechanism* 4, no. 3: 131–140. https://doi.org/10.1016/j.ddmec.2007.12.005

Hartley, L., M. Dyakova, J. Holmes, A. Clarke, M.S. Lee, E. Ernst, and K. Rees. 2014. Yoga for the primary prevention of cardiovascular disease. *Cochrane Database Systematic Review* 13, no. 5: CD010072. https://doi.org/10.1002/14651858.CD010072.pub2

He, F.J., and G.A. MacGregor. 2002. Effect of modest salt reduction on blood pressure: A meta-analysis of randomized trials. Implications for public health. *Journal of Human Hypertension* 16, no. 11: 761–770. https://doi.org/10.1038/sj.jhh.1001459

He, F.J., N.R. Campbell, and G.A. MacGregor. 2012. Reducing salt intake to prevent hypertension and cardio-vascular disease. *Revista Panamericana De Salud Publica* 32, no. 4: 293–300. https://doi.org/10.1590/s1020-49892012001000008

Howard, B.V., and W.R. Wylie-Rosett. 2002. Sugar and cardiovascular disease. *Circulation* 106, no. 4: 523–527. https://doi.org/10.1161/01.cir.0000019552.77778.04

Ittaman, S.V., J.J. VanWormer, and S.H. Rezkalla. 2014. The role of aspirin in the prevention of cardiovascular disease. *Clinical Medical Research* 12, no. 3–4: 147–154. https://doi.org/10.3121/cmr.2013.1197

Kale, J.S., R.R. Deshpande, and N.T. Katole. 2016. The effect of Sudarshan Kriya Yoga (SKY) on cardio-vascular and respiratory parameters. *International Journal of Medical Science and Public Health* 5: 2091–2094.

Katzel, L.I., and W. Shari. 2001. Classification of cardiovascular disease. In *Neuropsychology of Cardiovascular Disease*, edited by S.R. Waldstein and M.F. Elias. New York, 384pp. https://doi.org/10.4324/9781410600981

Kim, Y., Y. Je, and E.L. Giovannucci. 2021. Association between dietary fat intake and mortality from all-causes, cardiovascular disease, and cancer: A systematic review and meta-analysis of prospective cohort studies. *Clinical Nutrition* 40, no. 3: 1060–1070. https://doi.org/10.1016/j.clnu.2020.07.007

Kohl, H.W. 2002. Physical activity and cardiovascular disease: Evidence for a dose response. *Medicine and Science in Sports and Exercise* 33: S472–83; discussion S493–4. https://doi.org/10.1097/00005768-200106001-00017

Kondo, T., Y. Nakano, S. Adachi, and T. Murohara. 2019. Effects of tobacco smoking on cardiovascular dis-ease. *Circulation Journal* 83, no. 10: 1980–1985. https://doi.org/10.1253/circj.CJ-19-0323

Leon, B.M., and T.M. Maddox. 2015. Diabetes and cardiovascular disease: Epidemiology, biological mecha-nisms, treatment recommendations and future research. *World Journal of Diabetes* 6, no. 13: 1246–1258. https://doi.org/10.4239/wjd.v6.i13.1246

Lichtenstein, A.H., J. Lawrence, M.V. Appel, B.H. Frank, M. Penny, E. Kris, M.R. Casey, M.S. Frank, N.T. Anne, V.H. Linda, and W.-R. Judith. 2021. 2021 dietary guidance to improve cardiovascular health: A scientific statement from the American Heart Association. *Circulation* 144, no. 23. https://doi.org/10.1161/cir.0000000000001031

Llanaj, E., G.M. Dejanovic, E. Valido, A. Bano, M. Gamba, L. Kastrati, B. Minder, S. Stojic, T. Voortman, P. Marques-Vidal, J. Stoyanov, B. Metzger, M. Glisic, H. Kern, and T. Muka. 2022. Effect of oat supple-mentation interventions on cardiovascular disease risk markers: A systematic review and meta-analysis of randomized controlled trials. *European Journal of Nutrition* 61, no. 4: 1749–1778. https://doi.org/10.1007/s00394-021-02763-1

Lorena, S.P., V.L. James, E.M. Maria, L. Hector, G.A. Maria, D.S. Dorothy, A.T. Gregory, and A.M.M. Cheryl. 2020. Sugar-sweetened beverage intake and cardiovascular disease risk in the California teachers study. *Journal of the American Heart Association* 9, no. 10: e014883. https://doi.org/10.1161/JAHA.119.014883

Nezami, B.T., D.S. Ward, L.A. Lytle, S.T. Ennett, and D.F. Tate. 2018. A mHealth randomized controlled trial to reduce sugar-sweetened beverage intake in preschool-aged children. *Pediatric Obesity* 13, no. 11: 668–676. https://doi.org/10.1111/ijpo.12258

Nisar, A., S. Hasan, R. Goel, and L. Chaudhary. 2016. Impact of Sudarshan Kriya yoga on mean arterial blood pressure and biochemical parameters in medical students. *International Journal of Research in Medical Sciences*: 2150–2152. https://doi.org/10.18203/2320-6012.ijrms20161777

O'Donnell, M., A. Mente, M.H. Alderman, A.J.B. Brady, R. Diaz, R. Gupta, P. López-Jaramillo, F.C. Luft, T.F. Lüscher, G. Mancia, J.F.E. Mann, D. McCarron, M. McKee, F.H. Messerli, L.L. Moore, J. Narula, S. Oparil, M. Packer, D. Prabhakaran, A. Schutte, K. Sliwa, J.A. Staessen, C. Yancy, and S. Yusuf. 2020. Salt and cardiovascular disease: Insufficient evidence to recommend low sodium intake. *European Heart Journal* 41, no. 35: 3363–3373. https://doi.org/10.1093/eurheartj/ehaa586

Oyebode, O., V. Gordon-Dseagu, A. Walker, and J.S. Mindell. 2014. Fruit and vegetable consumption and all-cause, cancer and CVD mortality: Analysis of Health Survey for England data. *Journal of Epidemiology and Community Health* 68, no. 9: 856–862. https://doi.org/10.1136/jech-2013-203500

Ozen, E., R. Mihaylova, M. Weech, S. Kinsella, J.A. Lovegrove, and K.G. Jackson. 2022. Association between dietary saturated fat with cardiovascular disease risk markers and body composition in healthy adults: Findings from the cross-sectional BODYCON study. *Nutrition and Metabolism* 19, no. 1: 15. https://doi.org/10.1186/s12986-022-00650-y

Parajuli, N., B. Pradhan, and M. Jat. 2021. Effect of four weeks of integrated yoga intervention on perceived stress and sleep quality among female nursing professionals working at a tertiary care hospital: A pilot study. *Indian Psychiatry Journal* 30, no. 1: 136–140. https://doi.org/10.4103/ipj.ipj_11_21

Prabhakaran, D., P. Jeemon, and A. Roy. 2016. Cardiovascular diseases in India: Current epidemiology and future directions. *Circulation* 133, no. 16: 1605–1620. https://doi.org/10.1161/CIRCULATIONAHA.114.008729

Prateek, Y., K. Chatterjee, J. Prakash, N. Salhotra, V. Chauhan, and K. Srivastava. 2021. Impact of breathing and relaxation training (Sudarshan Kriya) on cases of alcohol dependence syndrome. *Industrial Psychiatry Journal* 30, no. 2: 341–345. https://doi.org/10.4103/ipj.ipj_117_21

Rippe, J.M. 2019. Lifestyle strategies for risk factor reduction, prevention, and treatment of cardiovascular disease. *American Journal of Lifestyle Medicine* 13, no. 2: 204–212. https://doi.org/10.1177/1559827618812395

Rippe, J.M., and T.J. Angelopoulos. 2016. Relationship between added sugars consumption and chronic disease risk factors: Current understanding. *Nutrients* 8, no. 11: 697. https://doi.org/10.3390/nu8110697

Rissanen, T.H., S. Voutilainen, J.K. Virtanen, B. Venho, M. Vanharanta, J. Mursu, and J.T. Salonen. 2003. Low intake of fruits, berries and vegetables is associated with excess mortality in men: The Kuopio Ischaemic Heart Disease Risk Factor (KIHD) study. *The Journal of Nutrition* 133, no. 1: 199–204. https://doi.org/10.1093/jn/133.1.199

Roth, G.A., C. Johnson, A. Abajobir, F. Abd-Allah, S.F. Abera, G. Abyu, M. Ahmed, B. Aksut, T. Alam, K. Alam, et al. 2017. Global, regional, and national burden of cardiovascular diseases for 10 causes, 1990 to 2015. *Journal of American College of Cardiology* 70, no. 1: 1–25. https://doi.org/10.1016/j.jacc.2017.04.052

Saladini, F., E. Benetti, C. Fania, L. Mos, E. Casiglia, and P. Palatini. 2016. Effects of smoking on central blood pressure and pressure amplification in hypertension of the young. *Vascular Medicine* 21, no. 5: 422–428. https://doi.org/10.1177/1358863X16647509

Sawicki, C.M., P.F. Jacques, A.H. Lichtenstein, G.T. Rogers, J. Ma, E. Saltzman, and N.M. McKeown. 2021. Whole- and refined-grain consumption and longitudinal changes in cardiometabolic risk factors in the Framingham offspring cohort. *Journal of Nutrition* 151, no. 9: 2790–2799. https://doi.org/10.1093/jn/nxab177

Sesso, H.D., R.S. Paffenbarger, and I.M. Lee. 2000. Physical activity and coronary heart disease in men: The Harvard Alumni Health Study. *Circulation* 102, no. 9: 975–980. https://doi.org/10.1161/01.cir.102.9.975

Soulaidopoulos, S., E. Nikiphorou, T. Dimitroulas, and G.D. Kitas. 2018. The role of statins in disease modification and cardiovascular risk in rheumatoid arthritis. *Frontiers in Medicine* 5: 24. https://doi.org/10.3389/fmed.2018.00024

Sreeniwas Kumar, A., and N. Sinha. 2020. Cardiovascular disease in India: A 360 degree overview. *Medical Journal Armed Forces India* 76, no. 1: 1–3. https://doi.org/10.1016/j.mjafi.2019.12.005

Sriram, K., H.S. Mulder, H.R. Frank, T.S. Santanam, A.C. Skinner, E.M. Perrin, S.C. Armstrong, E.D. Peterson, M.J. Pencina, and C.A. Wong. 2021. The dose-response relationship between physical activity and cardiometabolic health in adolescents. *American Journal of Preventive Medicine* 60, no. 1: 95–103. https://doi.org/10.1016/j.amepre.2020.06.027

Strazzullo, P., L. D'Elia, N.B. Kandala, and F.P. Cappuccio. 2009. Salt intake, stroke, and cardiovascular disease: Meta-analysis of prospective studies. *BMJ* 339: b4567. https://doi.org/10.1136/bmj.b4567

Tilman, D., and M. Clark. 2014. Global diets link environmental sustainability and human health. *Nature* 515, no. 7528: 518–522. https://doi.org/10.1038/nature13959

Tsao, C.W., and R.S. Vasan. 2015. Cohort profile: The Framingham Heart Study (FHS): Overview of milestones in cardiovascular epidemiology. *International Journal of Epidemiology* 44, no. 6: 1800–1813. https://doi.org/10.1093/ije/dyv337

Vaidyanathan, G. 2021. Healthy diets for people and the planet. *Nature* 600: 22–25.

Vanessa, S., M. Gregorio, D. Villafanha, and M. Godoy Moacir. 2018. Effects of intake of processed quinoa seeds on lipid profile in patients with coronary heart disease. *International Journal of Sciences* 4: 8–14. https://doi.org/10.18483/ijSci.1572

Wang, J., F. Liu, J. Li, K. Huang, X. Yang, J. Chen, X. Liu, J. Cao, S. Chen, C. Shen, L. Yu, F. Lu, X. Wu, L. Zhao, Y. Li, D. Hu, J. Huang, D. Gu, and X. Lu. 2022. Fruit and vegetable consumption, cardiovascular disease, and all-cause mortality in China. *Science China Life Sciences* 65, no. 1: 119–128. https://doi .org/10.1007/s11427-020-1896-x

Wang, M., H. Ma, Q. Song, et al. 2022. Red meat consumption and all-cause and cardiovascular mortality: Results from the UK Biobank study. *European Journal of Nutrition* 61: 2543–2553. https://doi.org/10 .1007/s00394-022-02807-0

Wang, Y., B. Chen, Y. Li, L. Zhang, Y. Wang, S. Yang, X. Xiao, and Q. Qin. 2021. The use of renin-angioten-sin-aldosterone system (RAAS) inhibitors is associated with a lower risk of mortality in hypertensive COVID-19 patients: A systematic review and meta-analysis. *Journal of Medical Virology* 93, no. 3: 1370–1377. https://doi.org/10.1002/jmv.26625

Wang, Y.J., T.L. Yeh, M.C. Shih, Y.K. Tu, and K.L Chien. 2020. Dietary sodium intake and risk of cardiovascular disease: A systematic review and dose-response meta-analysis. *Nutrients* 12: 2934. https://doi.org /10.3390/nu12102934

Wannamethee, S.G., and A.G. Shaper. 2001. Physical activity in the prevention of cardiovascular disease: An epidemiological perspective. *Sports Medicine* 31, no. 2: 101–114. https://doi.org/10.2165/00007256 -200131020-00003

Willett, W., J. Rockström, B. Loken, M. Springmann, T. Lang, S. Vermeulen, et al. 2019. Food in the anthropocene: The EAT–Lancet commission on healthy diets from sustainable food systems. *The Lancet*, no. 18: 31788–31784. https://doi.org/10.1016/s0140-6736

World Health Statistics 2022: Monitoring Health for the SDGs, Sustainable Development Goals. Geneva: World Health Organization, 2022. Licence: CC BY-NC-SA 3.0 IGO.

Yang, Q., Z. Zhang, E.W. Gregg, W.D. Flanders, R. Merritt, and F.B. Hu. 2014. Added sugar intake and cardiovascular diseases mortality among US adults. *JAMA Internal Medicine* 174, no. 4: 516–524. https:// doi.org/10.1001/jamainternmed.2013.13563

Ziaeian, B., and G.C. Fonarow. 2017. Statins and the prevention of heart disease. *JAMA Cardiology* 2, no. 4: 464. https://doi.org/10.1001/jamacardio.2016.4320

7 Cancer

7.1 INTRODUCTION

Cancer is a general term used for a group of diseases that are caused by various reasons including exposure to physical carcinogens such as radiation, chemical carcinogens such as aflatoxin contamination, smoking, alcoholism, asbestos, etc., and biological carcinogens such as pathogen infections leading to uncontrolled growth and division of abnormal cells. Cancer is a leading cause of death globally, and in 2020, it was estimated that there were 19.3 million new cases and nearly 10 million deaths (Sung et al. 2021) from various cancers including lung cancer (1.8 million), colon and rectum cancer (916,000), liver cancer (830,000), stomach cancer (769,000), breast cancer (685,000) (http://www.who.int/news-room/fact-sheets/detail/cancer). In terms of new cases of cancer, the most common cancers in 2020 were breast cancer (2.26 million), lung (2.21 million), colon and rectum (1.93 million), prostate (1.41 million), skin (non-melanoma) (1.2 million), and stomach (1.09 million) (Sung et al. 2021; http://www.who.int/news-room/fact-sheets/detail/cancer).

7.2 TYPES OF CANCERS

According to the National Cancer Institute (NIH), the common cancers are bladder cancer, breast cancer, colon and rectal cancer, endometrial cancer, kidney cancer, leukaemia, liver cancer, lung cancer, melanoma, non-Hodgkin lymphoma, pancreatic cancer, and thyroid cancer (http://www.cancer.gov/types) (Figure 7.1).

7.3 PREVALENCE

For men and women combined, 58.3% of cancer-related deaths occurred in Asia in 2020 (Sung et al. 2021). In men, lung cancer is the leading cause of mortality, followed by prostate and liver cancers. In women, breast and cervical cancer are the two leading causes of mortality. The age-standardised incidence rate per 100,000 individuals is provided in Figure 7.2, which shows the highest incidence of ≥257.1 in Australia, North America, and Europe. The number of individuals with cancer in India was projected to be 1,392,179 in 2020 with breast, lung, mouth, cervix uteri, and tongue as common cancer sites (Mathur et al. 2020). The northeast regions of India, particularly Aizawl and Papumpare districts, had the highest age-adjusted cancer incidence rate. The northeast states, followed by central and western states in India, had a higher proportion of cancer that is associated with tobacco usage. In Meghalaya, East Khasi Hills district, tobacco usage was found to be associated with the major proportion of cancer both in men (70.4%) and women (46.5%) (Mathur et al. 2020).

In Australia, among the 20 most common cancers that were diagnosed in 2021, prostate cancer, melanoma of the skin, colorectal cancer, lung cancer, and non-Hodgkin lymphoma occupied the top five cancers in men. On the other hand, the top five cancers in women were breast cancer, colorectal cancer, melanoma of the skin, lung cancer, and uterine cancer. However, lung cancer is the most common cancer, which was the number one reason for cancer-related deaths in 2021 among both men and women, followed by prostate cancer in men and breast cancer in women (Cancer in Australia 2021).

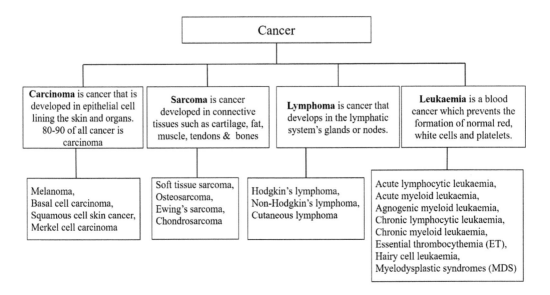

FIGURE 7.1   Common types of cancer.

7.4   AETIOLOGY

Aetiology of various cancers is described below; in which some are modifiable causative factors.

7.4.1   RADIATION

Ultraviolet radiation (UVR) and indoor tanning devices are classified as class I carcinogens by International Agency for Research on Cancer (IARC), which cause nearly 95% of all skin melanoma cases and related deaths in the US. The systematic review and meta-analysis conducted from 19 studies with 28,314 participants to understand the association between exposure to solar UVR and melanoma. The analysis revealed that there was an increased incidence of melanoma with a relative risk of 1.16, when there was occupational exposure to solar UVR compared to low or no occupational exposure to solar UVR. This increased risk was estimated to be 16% when followed over a lifetime (WHO 2021).

The sun protection survey conducted in Australia from 2016 to 2017 reported that 11% of adults were actively trying to get a tan while 66% of adults reported having tanned skin. Wearing a hat, sunscreen and other methods such as staying in shade, proper clothing to cover arms or legs could reduce the exposure to solar UVR (Cancer in Australia 2021) and therefore, it is a modifiable risk which requires behaviour change.

7.4.2   ROLE OF ALCOHOL

It is a modifiable risk factor for cancer. Metabolites of alcohol cause DNA damage that leads to cancer (Rumgay et al. 2021) and alcohol intake accounts for 4% of cancer worldwide. Alcohol consumption increases the risk of liver cancer, colorectal cancer, aerodigestive tract cancer, and breast cancer. Consuming alcohol is an independent risk factor for primary liver cancer (Baan et al. 2007). The meta-analysis conducted using cohort, case-control, and clinical studies showed that alcohol consumption was associated with liver cancer and related deaths. The clinical cohort studies showed that the history of ever drinking was a 1.17-fold increased risk of chronic liver diseases than never or low-frequency drinker (Chuang et al. 2015). Mortality from liver cancer was higher among people who have alcohol-related disorders, including abusers, alcoholism, and alcoholic cirrhosis, compared to the control group (Chuang et al. 2015). Former alcohol consumption was also shown to

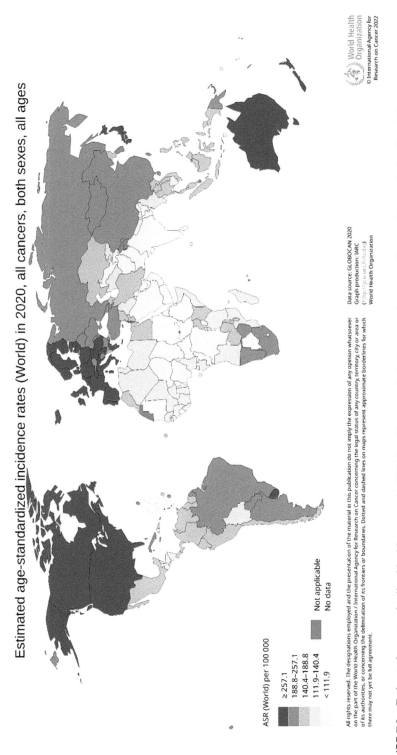

Estimated age-standardized incidence rates (World) in 2020, all cancers, both sexes, all ages

ASR (World) per 100 000

≥ 257.1
188.8–257.1
140.4–188.8
111.9–140.4
< 111.9
Not applicable
No data

Data source: GLOBOCAN 2020
Graph production: IARC
(http://gco.iarc.fr/today)
World Health Organization

World Health
Organization

© International Agency for
Research on Cancer 2022

FIGURE 7.2 Estimated age-standardized incidence rate in 2020 for all cancers in both men and women. (Source: Cancer Today (iarc.fr)).

be associated with an increased risk of liver cancer (mRR = 2.06) (Chuang et al. 2015). The study also revealed an increased relative risk with an increase in the amount of alcohol consumption. The relative risk was 1.08 for 12 g/day consumption, 1.54 with 50 g/day consumption, 2.14 with 75 g/day consumption, and 3.21 for 100 g/day consumption and 5.2 for 125 g/day consumption (Chuang et al. 2015). Another dose–response meta-analysis conducted using prospective studies using 16 published articles with a total of 4445 incident cases and 5550 deaths from liver cancer showed that the relative risk of consuming alcohol moderately (<3 drinks/day) was 0.91 compared to heavy drinkers (≥3 drinks/day) was 1.16 (Turati et al. 2014). The excessive risk was estimated to be 46% for 50 g of ethanol/day and 66% for 100 g/day (Turati et al. 2014).

In the population-based cohort study conducted among individuals with a lifetime of alcohol drinking as reported by themselves, it was found that alcohol high consumption of 2 to <3 times/day (hazard risk (HR)= 1.10) and very high of 3+ drinks/day (HR = 1.21) had a higher risk of overall mortality and combined risk of cancer or death compared to light alcohol drinkers for 1–3 drinks/week (HR= 1.08) and never or infrequent drinker for <1 drink/week (HR = 1.09) (Kunzmann et al. 2018).

The meta-analysis conducted using case-control studies showed that alcohol consumption was a significant risk factor for oesophageal cancer (odds ratio (OR) = 1.79), gastric cancer (OR = 1.4), hepatocellular carcinoma (OR = 1.56), nasopharyngeal cancer (OR = 1.21), and oral cancer (OR = 1.71). However, it did not have any significant risk factor for breast and gall bladder cancer. The pooled analyses of case-control and cohort studies showed that alcohol consumption was significantly associated with oesophageal cancer and gastric cancer with 1.78- and 1.4-fold higher risk, respectively (Li et al. 2011).

Analysis of data from 58,515 women with invasive breast cancer and 95,067 controls from 53 studies revealed that the relative risk of breast cancer was 1.32 among alcohol-consuming individuals who consume 35–44 g per day. The relative risk of breast cancer was even higher in individuals who consume ≥45 g of alcohol per day. For each additional 10 g increase in alcohol intake, there was 7.1% increase in the relative risk of breast cancer (Hamajima et al. 2002).

7.4.3 ROLE OF SMOKING

Smoking is one of the leading causes of preventable death. The retrospective study analysed the association between tobacco smoking and cancer using Cox's regression model in the UK and showed a positive association between smoking and several cancers such as liver, bladder, kidney, and pancreatic cancer, and lymphoma (Jacob et al. 2018). Another study on cigarette smoking and its association with colorectal cancer (Huang et al. 2022) showed that patients smoking 11–20 cigarettes per day or smoking for more than 30 years had a higher hazard risk of 1.16 and 1.14 respectively (Huang et al. 2022). The cohort study was conducted by following 96,855 Korean adult participants for a median 5.9 years, among whom 1250 participants developed thyroid cancer. The results showed smoking decreased the risk of thyroid cancer incidence in men but not in women, and this was consistent even after adjusting for thyroid-stimulating hormone and body mass index (BMI). The hazard risk was 0.58 for current and former smokers and 0.93 for never smokers (Cho et al. 2018).

7.4.4 ROLE OF AFLATOXIN

Aflatoxin B_1 (AFB_1) is a class I carcinogen that enters the body through contaminated food. Once ingested, AFB_1 is metabolised by the cytochrome P-450 system and forms an intermediate called AFB_1-8,9 epoxide, which can bind with DNA to form AFB_1-guanine adduct which is normally excreted through urine, and to protein to form AFB_1-albumin adduct which is normally found in blood (Wild and Turner 2002). In chronic exposure, the accumulation of aflatoxin in the liver causes liver cirrhosis and hepatocellular carcinoma (Williams et al. 2004). The community-based cohort study with molecular dosimetry of aflatoxin exposure conducted in the Penghu Islets in Taiwan, where hepatocellular carcinoma (HCC)-related mortality rate is highest, showed there was 65% (13/20) of HCC patients and 37% (32/86) of matched healthy controls were positive for aflatoxin-albumin adduct, and therefore the study concluded that the elevated risk of HCC among the Penghu residents may be due to their aflatoxin exposure along with hepatitis B surface antigen (HBsAg) carrier (Chen et al. 1996). The analysis of prospective studies which had evidence from combined food frequency questionnaires and food surveys of AFB_1 contamination, collecting food from plates to test the AFB_1 ingestion levels, urinary excretion of AFB_1 metabolites, and excretion of DNA adducts, showed a strong association between AFB_1 exposure in serum/urine and HCC, and the study concluded that dietary aflatoxin exposure is a contributor to the HCC incidents that occur in developing countries of Asia and sub-Saharan Africa, where 82% of cases occur (Wu and Santella 2012). The study to estimate the risk of aflatoxin-induced liver cancer used an epidemiological study conducted in three villages in Tanzania (Kimanya et al. 2021), which was considered representative of aflatoxin exposure in Tanzania. The data were used to assess the aflatoxin-induced liver cancer using disability-adjusted life years (DALYs). The exposure ranged from 15 to 10,926 ng/kg bw/day, and there were 1480 new cases of aflatoxin-induced liver cancer in 2016 (Kimanya et al. 2021).

7.5 TREATMENT

There are various treatments available to treat and/or cure cancer. These treatments are used depending on the stage of the cancer. The treatment begins with the diagnosis. Biomarker testing determines the genes, proteins, and other cancer biomarkers that provide information about the cancer. Chemotherapy uses drugs to kill the cancer cells. Hormone therapy slows the growth of cancers such as breast and prostate cancers. Hyperthermia is a type of treatment in which the body tissue is heated to 45°C and kills the cells. Immunotherapy helps the immune system fight cancer. Photodynamic therapy uses drugs to kill the cancer cells. Radiation therapy uses a high dosage of radiation to kill the cancer cells. Stem cell transplant restores the stem cells that are usually destroyed during chemo or radiation therapy. Surgery is used to treat cancer by removing the cancer from the body. Targeted therapy is a type of treatment that targets the changes in cancer cells that help them grow, divide and spread (Types of Cancer Treatment – NCI).

7.5.1 Dietary Modification

Early detection of cancer is key for providing appropriate treatment at the right time to save patients from cancer-related mortality and to cure some cancers. Between 30 and 50% of cancers can be prevented by avoiding preventable risk factors.

7.5.1.1 Reduce Saturated Fat and Trans-fat

An observational prospective study conducted from 12 cohorts shows there was a positive association between saturated fat consumption and increased risk of mortality from cancer (Kim et al. 2020). Results from 11 prospective cohorts show that the risk of mortality was low with high intake of polyunsaturated fatty acid (Kim et al. 2020).

7.5.1.2 Reduce Red Meat Consumption

A population-based cohort study called the Golestan cohort study, conducted in Iran among 50,045 individuals between the ages of 40 and 75, aimed at finding the association between meat consumption and oesophageal and gastric cancer. The study showed that there is an increased hazard risk among people who consume more red meat (HR = 1.08), which is red, processed, and organ meat, for gastric cancer (Collatuzzo et al. 2022). Similarly, the red meat intake (HR = 1.09) intake was also associated with gastric cancer. Red meat consumption was also found to be associated with oesophageal cancer among women, with high hazard risk of 1.13 for a one quintile increase in red meat consumption (Collatuzzo et al. 2022).

7.5.1.3 Increase Fruit and Vegetables

The health survey conducted in England used data collected among adults of >35 years adults was analysed to see the relationship between (Oyebode et al. 2014) fruit and vegetable consumption and the risk of cancer mortality (Oyebode et al. 2014). The result shows that increased consumption of fruit and vegetables was associated with decreased cancer-related mortality, vegetable consumption in particular was found to be more protective.

The dose–response meta-analysis of prospective studies using 58 studies from 52 articles showed the relationship between fruit and vegetable consumption on cancer. It shows there was a reduction in the relative risk of cancer with increased consumption of green-yellow vegetables and cruciferous vegetable consumption up to 600 g per day, and the relationship was non-linear after 600 g per day (Aune et al. 2017). There were 14%, 8%, and 12% reductions in the relative risk with the intakes of 550–600 g/day of fruit and vegetables., fruit alone, and vegetables alone, respectively.

The study conducted by reviewing around 250 observational studies in Dutch estimated that an increase in the consumption of fruit and vegetables. by 150 g/day from 250 g/day (in total 400 g/day) will reduce cancer incidence by 19% (with the best guess method), 6% (with the conservative method), and 28% (with the optimistic method). This would result in 12,000, 4500, and 17,500 preventable cases estimated with the best guess, conservative, and optimistic methods (Vant's Veer et al. 1999).

The systematic review and meta-analysis of ten cohort studies determined the mean relative risk (RR) of low consumption of fruit is 0.85, with the range of 0.55 to 1.92, for stomach cancer. In 28 cohort studies, the same risk was determined to be 0.63, with the range of 0.31 to 1.39 (Vainio and Weiderpass 2009). On the other hand, the relative risk of low vegetable consumption on stomach cancer was 0.94, with the range of 0.7 to 1.25, which was evaluated from five cohort studies. From 20 case-control studies, it was evaluated as 0.66, with the range of 0.3 to 1.7 (Vainio and Weiderpass 2009).

Another systematic review and meta-analysis of five cohort studies reported that high vegetable consumption is associated with less overall mortality in survivors of head and neck cancer with a relative HR of 0.75. Four meta-analyses showed similar inverse association with ovarian cancer-related mortality with an HR of 0.82 (Hurtado-Barroso et al. 2020). There was no association found in other cancers such as colorectal cancer and respiratory tract cancer.

7.5.1.4 Increase Wholegrain Consumption.

The population-based Swedish mammography cohort study conducted through prospective examination of the association between wholegrain consumption and colorectal cancer revealed that high consumption of wholegrain ($\geq$4.5 servings a day) reduced the relative risk of colon cancer (RR = 0.65) compared to low wholegrain consumers (<1.5 servings a day) but no association was found for rectal cancer (Larsson et al. 2005). However, another study conducted among Norwegian women to determine the relationship between wholegrain consumption and colorectal cancer showed that there was only a weak association between wholegrain consumption and proximal rectal cancer, and not distal rectal cancer and rectal cancer (Bakken et al. 2016). The dose–response systematic review and meta-analysis conducted from six cohort studies showed that the high intake of wholegrains reduces the relative risk (RR = 0.89) of total cancer, and the summary relative risk of cancer-related mortality per 90 g/day intake was 0.85 (Aune et al. 2016a). The systematic review and meta-analysis conducted to determine the association between wholegrain consumption and cancer risk from seven studies showed that wholegrain consumption reduced the risk of total cancer by 6–12%. This reduction was 3–20% if the dosage is between 15 and 90 g of wholegrain per day. A consistent association was found in colorectal, colon, gastric, pancreatic, and oesophageal cancers. In contrast, high consumption of refined grain increased the risk of colon and gastric cancers, however, the result is inconclusive, due to the fact that there are only limited studies on this and the reason that refined grain was often defined to include both staple grain foods and indulgent grain foods (Gaesser 2020). The study conducted to determine the association between wholegrain consumption and the risk of gastric cancer showed that there was a 44% reduction in the risk of gastric cancer with a large level of wholegrain consumption (Wang et al. 2020). In contrast, refined grain consumption increased the risk by 36%, and this association was even higher with a large amount of refined grain, which increased the risk by 63% compared to moderate amounts of refined grain consumption, which increased the risk by 28% (Wang et al. 2020).

7.5.1.5 Increase Nut Consumption

A study conducted among women with pancreatic cancer showed that women who consumed 28 g of nuts more than two times per week demonstrated a reduced relative risk of pancreatic cancer (RR = 0.65) compared to non-consumers (Bao et al. 2013) and this inverse association was independent of other potential risks for pancreatic cancer such as physical activity, BMI, smoking, intake of red meat, and low fruit and vegetable consumption. The dose–response meta-analysis conducted using nine cohort studies from eight published articles to determine the association between nut consumption and relative risk of cancer (Aune et al. 2016b) showed that the relative risk of cancer for high and low consumption of nuts was 0.82, with a summary relative risk of cancer for one serving of nuts per a day of 0.85. The high vs low consumption of tree nuts and peanuts had a summary relative risk of cancer of 0.82 and 0.93, respectively. This risk further decreased by 10 g/day, to 0.80 and 0.92 for peanuts respectively (Aune et al. 2016b). The association between nut consumption and the risk of cancer was studied through a systematic review and a dose–response meta-analysis of prospective studies (Long et al. 2020) revealed that high consumption of nuts is associated with a reduced relative risk of overall cancer (RR = 0.90), and this effect was apparent only in digestive system cancer with a relative risk of 0.83. Among various nuts, tree nuts were found to be associated with a reduced relative risk (0.86) of cancer. Eighteen prospective studies were used to conduct a dose–response analysis (Long et al. 2020) in which a significant effect on cancer was demonstrated when more than 9 g of nuts were consumed every day, with a relative risk of 0.95. An increase of 20 g/day in nut consumption was associated with a 10% reduced risk of cancer (RR = 0.90).

The above studies described the association between nut consumption and cancer risk; however, there are also studies that found no association between nut consumption and the risk of cancer. The meta-analysis of three big prospective cohort studies showed that there is no association between nut consumption compared to non-consumers and cancer, with the hazard risks of total cancer, 0.88 for lung cancer 0.88, 0.90 for bladder cancer, 0.96 for breast cancer, 1.07 for colorectal cancer, and 1.18 for prostate cancer among people who consumed nuts more than five times per week (Fang et al. 2021).

7.5.2 Physical Activity

The study pooled data from 12 prospective studies conducted in the US and European cohort was used to determine the association between physical activity and 26 different types of cancer through multivariant Cox's regression. The participants self-reported their physical activities, and the results showed a lower risk for 13 cancers with high levels of physical activity. The hazard risk was low for oesophageal adenocarcinoma (HR = 0.58), liver cancer (HR = 0.73), lung cancer (HR = 0.74), kidney cancer (HR = 0.77), gastric cardia (HR= 0.78), endometrial cancer (HR = 0.79), myeloid leukaemia (HR = 0.80), myeloma (HR = 0.83), colon (HR = 0.84), head and neck (HR = 0.85), rectal (HR = 0.87), bladder (HR = 0.87), and breast cancer (HR = 0.90) with high levels of physical activity. At least 10 out of all these 13 cancers had significant association even after being adjusted for BMI (Moore et al. 2016) with high physical activity. The systematic review and meta-analysis conducted to study the effect of physical activity in reducing the risk of cancer showed a 10–20% relative risk reduction with high physical activity in the case of lung, bladder, breast, colon, endometrial, oesophageal adenocarcinoma, renal, and gastric cancers. There was a significant risk reduction among people who had high physical activity compared to low physical activity, with relative risks of 0.85, 0.81, 0.80, 0.87, 0.88, 0.75, and 0.79 for bladder cancer, gastric cancer, colon cancer, endometrial cancer, breast cancer, renal cancer, lung cancer, and oesophageal cancer, respectively (McTiernan et al. 2019). The cancer-specific mortality was 40–50% less for breast, colorectal, or prostate cancer among individuals who engaged in greater physical activity, and the association was moderate or limited between cancer-specific mortality and physical activity (McTiernan et al. 2019). The self-reported physical activity to determine the association, moreover, most of the studies used for meta-analysis were from prospective studies. Modern devices are recommended to quantify the granular details of physical activity, including its intensity, duration, propensity, and frequency, to further understand how much physical activity is required to maintain health. It is also recommended to conduct randomised controlled studies and observational studies for quantification purpose (Schrack et al. 2017).

7.5.3 LIFESTYLE MODIFICATION

Lifestyle modification is important to prevent, reduce the risk, and also to manage the quality of life in cancer survivors. A modification in the diet in such a way that a healthy nutritious diet emphasising plant sources (LoConte et al. 2018) and made of plenty of fruit and vegetables, who-legrains, nuts and less saturated fat and red meat will help in managing the metabolic changes, obesity, diabetes, hyperlipidaemia in cancer patients or survivors which is very important to improve the quality of life. Using sunscreen, wearing protective clothing, limiting the exposure to the sun between 10 am and 2 pm, and avoiding indoor tanning also is very important to prevent skin cancer (LoConte et al. 2018). At the same time, other risk factors such as no smoking, and no alcohol consumption are also important to ensure the quality of life. Improving physical activity has proven benefits among cancer survivors to improve the quality of life by reducing other risk factors such as obesity, diabetes, and hyperlipidaemia. In a study, 64 hours of intensive nutrition, culinary medicine, physical activity, and stress relief practice were given over a six-month period of intervention to 57 cancer survivors of over 12 months (Golubić et al. 2018). The pre- and post-analysis of weight, BMI, waist circumference, blood pressure, other standard laboratory tests such as lipid profile, C-reactive protein, fasting insulin/glucose, and insulin resistance were conducted. The participants were also interviewed with a questionnaire to understand stress, depression, and quality of life. Hyperlipidaemia, hypertension, diabetes, prediabetes, obesity, overweight, and depression were diagnosed among 47%, 57%, 22%, 50%, 47%, 24%, and 16% of participants respectively. After the 12 months of intervention, BMI decreased significantly by an average of 2.4 kg/m^2, HDL level increased by 3.3 mg/dL, triglyceride level decreased by 23 mg/dL, C-reactive protein decreased by 1.3 mg/L, fasting insulin decreased by 4.2 μU/mL, and insulin resistance decreased by 1.5. There was also a significant improvement in perceived stress and quality of life. Employing lifestyle modification could produce benefits on health and quality of life (Golubić et al. 2018). A recent study reported that breast cancer survivors are at a higher cardiometabolic risk such as diabetes, high blood pressure, dyslipidaemia than women who have never been treated for breast cancer (Fillon 2022). A systematic review and meta-analysis was conducted (Lahart et al. 2018) to determine the effect of physical activity intervention on breast cancer treatment-related (adjuvant therapy) adverse physiological and psychological outcomes among 3239 intervention women against 2524 control group of women (who do not perform any physical activity). The physical activity was recorded through pre and post-measurement of self-reported physical activity, objectively measured with an accelerometer or pedometer. The results showed that there was a significant small to moderate improvement in health-related quality of life (standardised mean difference (SMD) 0.39), emotional function (SMD = 0.21), perceived physical function (SMD = 0.33), anxiety (SMD = −0.57), and cardiorespiratory fitness (SMD = 0.44). Physical activity is the most powerful lifestyle factor that can reduce breast cancer-related mortality by 40% (Lahart et al. 2018; Hamer et al. 2017). The review conducted on the effect of physical activity on breast cancer recommended 150 minutes of moderate to vigorous exercise or 75 minutes of vigorous exercise per week along with strength training for two to three weekly sessions (Hamer et al. 2017). One hundred and forty-one gynaecological malignancy patients who underwent surgery were evaluated to understand their lifestyle modification after surgery. The results showed that almost 89% of the patients changed their lifestyle after they were diagnosed with cancer. Sixty-six percent of them improved their nutrition by increasing consumption of fruit and vegetables, 65% of them ate less meat, 27% of them increased physical activity, 63% of them who were previous smokers quit or reduced it after diagnosis, 47% of them reduced alcohol consumption, and 77% of them described that their perceived stress level was reduced (Paepke et al. 2021). The study found more positive changes among the participants after cancer diagnosis shows that it could be a way to improve the quality of life and mental stability. However, the change in policy on some of the preventable risk factors could help in preventing the risk of cancer, which includes increasing alcohol taxes, reducing the number of alcohol outlets, limiting the exposure to alcohol marketing and advertisement among youth, improving screening methods, were shown to be a proven prevention strategy which could have high impact on reducing the development of melanoma and breast cancer, colorectal, head and neck, liver, and oesophageal cancer (LoConte et al. 2018).

7.6 CONCLUSION

Cancer is a life-threatening disease; however, if diagnosed early, there is a higher chance of providing treatment to cure it. Lifestyle, physical activity, and diet play a key role in developing cancer and increasing related risks in the body. Furthermore, lifestyle, physical activity, and diet are modifiable risk factors that help reduce the risk of developing cancer and improve the quality of life in cancer survivors.

REFERENCES

Aune, D., N. Keum, E. Giovannucci, L.T. Fadnes, P. Boffetta, D.C. Greenwood, S. Tonstad, L.J. Vatten, E. Riboli, and T. Norat. 2016a. Whole-grain consumption and risk of cardiovascular disease, cancer, and all-cause and cause-specific mortality: Systematic review and dose-response meta-analysis of prospective studies. *BMJ* 353: i2716. https://doi.org/10.1136/bmj.i2716:10.1136/bmj.i2716

Aune, D., N. Keum, E. Giovannucci, et al. 2016b. Nut consumption and risk of cardiovascular disease, total cancer, all-cause and cause-specific mortality: A systematic review and dose-response meta-analysis of prospective studies. *BMC Medicine* 14: 207. https://doi.org/10.1186/s12916-016-0730-3

Baan, R., K. Straif, Y. Grosse, B. Secretan, F. El Ghissassi, V. Bouvard, A. Altieri, and V. Cogliano. 2007. WHO International Agency for Research on Cancer Monograph Working Group. Carcinogenicity of alcoholic beverages. *Lancet Oncology* 8, no. 4: 292–293. PMID: 17431955. https://doi.org/10.1016/s1470-2045(07)70099-2

Bakken, T., T. Braaten, A. Olsen, C. Kyrø, E. Lund, and G. Skeie. 2016. Consumption of whole-grain bread and risk of colorectal cancer among Norwegian women (the NOWAC Study). *Nutrients* 8, no. 1: 40. https://doi.org/10.3390/nu8010040

Bao, Y., F.B. Hu, E.L. Giovannucci, B.M. Wolpin, M.J. Stampfer, W.C. Willett, and C.S. Fuchs. 2013. Nut consumption and risk of pancreatic cancer in women. *British Journal of Cancer* 109, no. 11: 2911–2916. https://doi.org/ 10.1038/bjc.2013.665

Cancer in Australia. 2021. Australian Institute of Health and Welfare 2021. Cancer in Australia 2021. Cancer series no. 133. Cat. no. CAN 144. Canberra: AIHW. Cancer in Australia 2021 (aihw.gov.au)

Chen, C.J., L.Y. Wang, S.N. Lu, M.H. Wu, S.L. You, Y.J. Zhang, L.W. Wang, and R.M. Santella. 1996. Elevated aflatoxin exposure and increased risk of hepatocellular carcinoma. *Hepatology* 24, no. 1: 38–42. PMID: 8707279. https://doi.org/10.1002/hep.510240108

Cho, A., Y. Chang, J. Ahn, H. Shin, and S. Ryu. 2018. Cigarette smoking and thyroid cancer risk: A cohort study. *British Journal of Cancer* 119, no. 5: 638–645. https://doi.org/10.1038/s41416-018-0224-5

Chuang, S.C., Y.C. Lee, G.J. Wu, K. Straif, and M. Hashibe. 2015. Alcohol consumption and liver cancer risk: A meta-analysis. *Cancer Causes Control* 26, no. 9: 1205–1231. https://doi.org/10.1007/s10552-015-0615-3

Collatuzzo, G., A. Etemadi, M. Sotoudeh, A. Nikmanesh, H. Poustchi, M. Khoshnia, A. Pourshams, M. Hashemian, G. Roshandel, S.M. Dawsey, C.C. Abnet, F. Kamangar, P. Brennan, P. Boffetta, and R. Malekzadeh. 2022. Meat consumption and risk of esophageal and gastric cancer in the Golestan Cohort Study, Iran. *International Journal of Cancer* 30. https://doi.org/10.1002/ijc.34056

Dagfinn, A., E. Giovannucci, P. Boffetta, L.T. Fadnes, N. Keum, T. Norat, D.C. Greenwood, E. Riboli, L.J. Vatten, and S. Tonstad. 2017. Fruit and vegetable intake and the risk of cardiovascular disease, total cancer and all-cause mortality—A systematic review and dose-response meta-analysis of prospective studies. *International Journal Of Epidemiology* 46, no. 3: 1029–1056. https://doi.org/10.1093/ije/dyw319

Fang, Z., Y. Wu, Y. Li, X. Zhang, W.C. Willett, A.H. Eliassen, B. Rosner, M. Song, L.A. Mucci, and E.L. Giovannucci. 2021. Association of nut consumption with risk of total cancer and 5 specific cancers: Evidence from 3 large prospective cohort studies. *American Journal of Clinical Nutrition* 114, no. 6: 1925–1935. https://doi.org/10.1093/ajcn/nqab295

Fillon, M. 2022. Breast cancer survivors face greater cardiometabolic risks. *CA: A Cancer Journal for Clinicians* 72: 303–304. https://doi.org/10.3322/caac.21746

Gaesser, G.A. 2020. Whole grains, refined grains, and cancer risk: A systematic review of meta-analyses of observational studies. *Nutrients* 12, no. 12: 3756. https://doi.org/10.3390/nu12123756

Golubić, M., D. Schneeberger, K. Kirkpatrick, J. Bar, A. Bernstein, F. Weems, J. Ehrman, J. Perko, J. Doyle, and M. Roizen. 2018. Comprehensive lifestyle modification intervention to improve chronic disease risk factors and quality of life in cancer survivors. *The Journal of Alternative and Complementary Medicine* 24, no. 11: 1085–1091. https://doi.org/10.1089/acm.2018.0193

Hamajima, N., K. Hirose, K. Tajima, T. Rohan, E.E. Calle, C.W. Heath, R.J. Coates, J.M. Liff, et al. 2002. Collaborative Group on Hormonal Factors in Breast Cancer. Alcohol, tobacco and breast cancer—Collaborative reanalysis of individual data from 53 epidemiological studies, including 58,515 women with breast cancer and 95,067 women without the disease. *British Journal of Cancer* 87, no. 11: 1234–1245. https://doi.org/10.1038/sj.bjc.6600596

Hamer, J., and E. Warner. 2017. Lifestyle modifications for patients with breast cancer to improve prognosis and optimize overall health. *CMAJ* 189, no. 7: E268–E274. https://doi.org/10.1503/cmaj.160464

Huang, Y.M., P.L. Wei, C.H. Ho, and C.C. Yeh. 2022. Cigarette smoking associated with colorectal cancer survival: A nationwide, population-based cohort study. *Journal of Clinical Medicine* 11, no. 4: 913. https://doi.org/10.3390/jcm11040913

Hurtado-Barroso, S., M. Trius-Soler, R.M. Lamuela-Raventós, and R. Zamora-Ros. 2020. Vegetable and fruit consumption and prognosis among cancer survivors: A systematic review and meta-analysis of cohort studies. *Advanced Nutrition* 11, no. 6: 1569–1582. PMID: 32717747; PMCID: PMC7666913. https://doi.org/10.1093/advances/nmaa082

Islami, F., A. Goding Sauer, K.D. Miller, R.L. Siegel, S.A. Fedewa, E.J. Jacobs, M.L. McCullough, A.V. Patel, J. Ma, I. Soerjomataram, W.D. Flanders, O.W. Brawley, S.M. Gapstur, and A. Jemal. 2018. Proportion and number of cancer cases and deaths attributable to potentially modifiable risk factors in the United States. *CA Cancer Journal for Clinicians* 68, no. 1: 31–54. https://doi.org/10.3322/caac.21440

Jacob, L., M. Freyn, M. Kalder, K. Dinas, and K. Kostev. 2018. Impact of tobacco smoking on the risk of developing 25 different cancers in the UK: A retrospective study of 422,010 patients followed for up to 30 years. *Oncotarget* 9, no. 25: 17420–17429. https://doi.org/10.18632/oncotarget.24724

Kim, Y., Y. Je, and E.L. Giovannucci. 2020. Association between dietary fat intake and mortality from all-causes, cardiovascular disease, and cancer: A systematic review and meta-analysis of prospective cohort studies. *Clinical Nutrition* 40, no. 3: 1060–1070. https://doi.org/10.1016/j.clnu.2020.07.007

Kimanya, M.E., M.N. Routledge, E. Mpolya, C.N. Ezekiel, C.P. Shirima, and Y.Y. Gong. 2021. Estimating the risk of aflatoxin-induced liver cancer in Tanzania based on biomarker data. *Plos One* 16, no. 3: e0247281. https://doi.org/10.1371/journal.pone.0247281

Kunzmann, A.T., H.G. Coleman, W.Y. Huang, and S.I. Berndt. 2018. The association of lifetime alcohol use with mortality and cancer risk in older adults: A cohort study. *PLoS Medicine* 15, no. 6: e1002585. https://doi.org/10.1371/journal.pmed.1002585

Lahart, I.M., G.S. Metsios, A.M. Nevill, and A.R. Carmichael. 2018. Physical activity for women with breast cancer after adjuvant therapy. *Cochrane Database Systematic Review* 1, no. 1: CD011292. https://doi.org/10.1002/14651858.CD011292.pub2

Larsson, S.C., E. Giovannucci, L. Bergkvist, and A. Wolk. 2005. Whole grain consumption and risk of colorectal cancer: A population-based cohort of 60,000 women. *British Journal of Cancer* 92, no. 9: 1803–1807. https://doi.org/10.1038/sj.bjc.6602543

Li, Y., H. Yang, and J. Cao. 2011. Association between alcohol consumption and cancers in the Chinese population–A systematic review and meta-analysis. *PLoS One* 6, no. 4: e18776. https://doi.org/10.1371/journal.pone.0018776

LoConte, N.K., C.A. Gershenwald, C. Thomson, T.E. Crane, G. Harmon, and R. Rechis. 2018. Lifestyle modifications and policy implications for primary and secondary cancer prevention on: Diet, exercise, sun safety, and alcohol reduction. *American Society of Clinical Oncology Educational Book* 38: 88–100. https://doi.org/10.1200/EDBK_200093

Long, J., Z. Ji, P. Yuan, T. Long, K. Liu, J. Li, and L. Cheng. 2020. Nut consumption and risk of cancer: A meta-analysis of prospective studies. *Cancer Epidemiology Biomarkers Prevention* 29, no. 3: 565–573. https://doi.org/10.1158/1055-9965.EPI-19-1167

Makarem, N., J.M. Nicholson, E.V. Bandera, N.M. McKeown, and N. Parekh. 2016. Consumption of whole grains and cereal fiber in relation to cancer risk: A systematic review of longitudinal studies. *Nutrition Review* 74, no. 6: 353–373. https://doi.org/10.1093/nutrit/nuw003

Mathur, P., K. Sathishkumar, M. Chaturvedi, P. Das, K.L. Sudarshan, S. Santhappan, V. Nallasamy, A. John, S. Narasimhan, and F.S. Roselind. 2020. ICMR-NCDIR-NCRP Investigator Group. Cancer statistics, 2020: Report from national cancer registry programme, India. *JCO Global Oncology* 6: 1063–1075. https://doi.org/10.1200/GO.20.00122

McTiernan, A., C.M. Friedenreich, P.T. Katzmarzyk, K.E. Powell, R. Macko, D. Buchner, L.S. Pescatello, B. Bloodgood, B. Tennant, A. Vaux-Bjerke, S.M. George, R.P. Troiano, and K.L. Piercy. 2019. Physical activity guidelines advisory committee. Physical Activity in Cancer Prevention and Survival: A Systematic Review. *Medicine and Science in Sports and Exercise* 51, no. 6: 1252–1261. https://doi.org/10.1249/MSS.0000000000001937

Moore, S.C., I.M. Lee, E. Weiderpass, P.T. Campbell, J.N. Sampson, C.M. Kitahara, S.K. Keadle, H. Arem, A. Berrington de Gonzalez, P. Hartge, H.O. Adami, C.K. Blair, K.B. Borch, E. Boyd, D.P. Check, A. Fournier, N.D. Freedman, M. Gunter, M. Johannson, K.T. Khaw, M.S. Linet, N. Orsini, Y. Park, E. Riboli, K. Robien, C. Schairer, H. Sesso, M. Spriggs, R. Van Dusen, A. Wolk, C.E. Matthews, and A.V. Patel. 2016. Association of leisure-time physical activity with risk of 26 types of cancer in 1.44 million adults. *JAMA Internal Medicine* 176, no. 6: 816–825. https://doi.org/10.1001/jamainternmed.2016.1548

Oyinlola, O., V. Gordon-Dseagu, A. Walker, and J.S. Mindell. 2014. Fruit and vegetable consumption and all-cause, cancer and CVD mortality: Analysis of health survey for England data. *Journal of Epidemiology and Community Health* 68, no. 9: 856–862. https://doi.org/10.1136/jech-2013-203500

Paepke, D., C. Wiedeck, A. Hapfelmeier, et al. 2021. Lifestyle modifications after the diagnosis of gynecological cancer. *BMC Women's Health* 21. https://doi.org/10.1186/s12905-021-01391-5

Rumgay, H., N. Murphy, P. Ferrari, and I. Soerjomataram. 2021. Alcohol and cancer: Epidemiology and biological mechanisms. *Nutrients* 13, no. 9: 3173. https://doi.org/10.3390/nu13093173

Schrack, J.A., G. Gresham, and A.A. Wanigatunga. 2017. Understanding physical activity in cancer patients and survivors: New methodology, new challenges, and new opportunities. *Cold Spring Harbor Molecular Case Studies* 3, no. 4: a001933. https://doi.org/10.1101/mcs.a001933

Sung, H., J. Ferlay, R.L. Siegel, M. Laversanne, I. Soerjomataram, A. Jemal, and B. Freddie. 2021. Global cancer statistics 2020: GLOBOCAN estimates of incidence and mortality worldwide for 36 cancers in 185 countries. *CA: A Cancer Journal for Clinicians*: 209–249. https://doi.org/10.3322/caac.21660

Turati, F., C. Galeone, M. Rota, C. Pelucchi, E. Negri, V. Bagnardi, G. Corrao, P. Boffetta, and C. La Vecchia. 2014. Alcohol and liver cancer: A systematic review and meta-analysis of prospective studies. *Annals of Oncology* 25, no. 8: 1526–1535. https://doi.org/10.1093/annonc/mdu020

Vainio, H., and E. Weiderpass. 2006. Fruit and vegetables in cancer prevention. *Nutrition and Cancer* 54, no. 1: 111–142. https://doi.org/10.1207/s15327914nc5401_13

van't Veer, P., M.C. Jansen, M. Klerk, and F.J. Kok. 2000. Fruits and vegetables in the prevention of cancer and cardiovascular disease. *Public Health Nutrition* 3, no. 1: 103–107. https://doi.org/10.1017/s1368980000000136

Wang, T., R. Zhan, J. Lu, L. Zhong, X. Peng, M. Wang, and S. Tang. 2020. Grain consumption and risk of gastric cancer: A meta-analysis. *International Journal of Food Science and Nutrition* 71, no. 2: 164–175. https://doi.org/10.1080/09637486.2019.1631264

WHO. 2021. The effect of occupational exposure to solar ultraviolet radiation on malignant skin melanoma and non-melanoma skin cancer: A systematic review and meta-analysis from the WHO/ILO Joint Estimates of the Work-related Burden of Disease and Injury. Geneva: World Health Organization. Licence: CC BY-NC-SA 3.0 IGO.

Wild, C.P., and P.C. Turner. 2002. The toxicology of aflatoxins as a basis for public health decisions. *Mutagenesis* 17, no. 6: 471–481. https://doi.org/10.1093/mutage/17.6.471

Williams, J.H., T.D. Phillips, P.E. Jolly, J.K. Stiles, C.M. Jolly, and D. Aggarwal. 2004. Human aflatoxicosis in developing countries: A review of toxicology, exposure, potential health consequences, and interventions. *Ammerican Journal of Clinical Nutrition* 80, no. 5: 1106–1122. PMID: 15531656. https://doi.org/10.1093/ajcn/80.5.1106

Wu, H.C., and R. Santella. 2012. The role of aflatoxins in hepatocellular carcinoma. *Hepatitis Monthly* 12, no. 10: e7238. https://doi.org/10.5812/hepatmon.7238

8 Feeding and Eating Disorders

8.1 INTRODUCTION

Eating disorders, formally known as feeding and eating disorders, are a type of serious and potentially fatal health conditions occurring due to severe disturbances in eating behaviours. Eating disorders are mental health conditions influenced by an obsession with food and body shape. The abnormal eating behaviour has a profound adverse effect on physical and mental states, and even social functioning. Many times, these fatal eating disorders go unnoticed and ignored in several societies, increasing the rate of psycho-physiological morbidity and mortality, thereby posing a huge health challenge..

8.2 CLASSIFICATION OF FEEDING AND EATING DISORDERS

Earlier, eating disorders were mainly classified into three categories, i.e., anorexia nervosa (AN), bulimia nervosa (BN), and other eating disorders that did not meet the criteria of AN and BN were categorised as eating disorders not otherwise specified (EDNOS). However, this classification was not enough to draw a factual clear picture of the eating disorders existing in the population, and therefore, it was felt necessary to reclassify eating disorders with a few amendments in the diagnostic criteria.

In 2013, the eating disorders categories were reclassified in the *Diagnostic and Statistical Manual of Mental Disorders* (DSM-5) as AN, BN, Binge Eating Disorder (BED), Other Specified Feeding and Eating Disorder (OSFED), Avoidant/Restrictive Food Intake Disorder (ARFID), Pica, Ruminant Disorder, Unspecified Feeding or Eating Disorder (UFED), and others like Muscle Dysmorphic Disorder and Orthorexia Nervosa (ON); the proposed criteria are shown in Figure 8.1 (APA 2013).

8.3 DIAGNOSIS AND ASSESSMENT CRITERIA

According to DSM-5 (APA 2013) criteria have been specified for diagnosis and assessment of each of the feeding and eating disorders which are as follows:

8.3.1 ANOREXIA NERVOSA (AN)

- Significantly low body weight due to restriction in energy intake
- Intense fear of gaining weight or becoming fat, even though at a significantly low weight
- Disturbance in the way in which one's body weight or shape is experienced, undue influence of body weight or shape on self-evaluation, or persistent lack of the seriousness of the current low body weight
- The level of severity is considered mild if BMI >17 kg/m^2, moderate if BMI is 16–16.99 kg/m^2, severe if BMI is 15–15.99 kg/m^2 and extreme if BMI is <15 kg/m^2.

8.3.2 BULIMIA NERVOSA (BN)

- Recurrent episodes of binge eating, i.e., eating a large amount of food in a discrete period of time and lack of control on what to eat and how much to eat

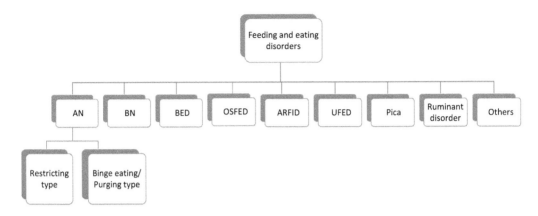

FIGURE 8.1   Classification of feeding and eating disorders.

- Recurrent inappropriate compensatory behaviours such as self-induced vomiting, fasting, misuse of laxatives, diuretics or medications, or excessive exercise in order to prevent weight gain
- Both the binge eating and inappropriate compensatory behaviours occur on average, at least once a week for three months
- Self-evaluation is influenced by body shape and weight
- The level of severity is considered mild when 1–3 episodes of inappropriate compensatory behaviour occur in a week, moderate when 4–7 episodes occur, severe when 8–13 episodes occur, and extreme when 14 or more episodes occur per week

8.3.3   BINGE EATING DISORDER (BED)

- BED is characterised by three or more of the following:
  - Eating more quickly than normal
  - Eating until feeling uncomfortably full
  - Eating a large amount even when not feeling physically hungry
  - Eating alone because of embarrassment from the amount of eating
  - Feeling disgusted with oneself, depressed or guilty afterwards
- Marked distress due to binge eating
- Binge eating occurs on an average at least once a week for three months
- Unlike BN, binge eating is not associated with the recurrent use of inappropriate compensatory behaviours
- The level of severity is considered mild if the binge eating episodes are 1–3 per week, moderate if 4–7, severe if 8–13, and extreme 14 or more episodes per week

8.3.4   OTHER SPECIFIED FEEDING AND EATING DISORDERS (OSFED)

In this category, those specific eating or feeding disorders are included that do not meet the criteria of any specific feeding and eating disorders such as atypical anorexia nervosa, bulimia nervosa of limited duration or low-frequency binge-eating disorder of limited duration or low-frequency purging disorder in the absence of binge eating, night eating syndrome.

8.3.5   PICA

Eating of non-nutritive or non-food substances over a period of at least one month.

8.3.6 Ruminant Disorder

Ruminant disorder is repeated regurgitation of food over a period of at least one month, which may be re-chewed, re-swallowed, or spat out.

8.3.7 Avoidant/Restrictive Food Intake Disorders (ARFID)

It is an eating disorder in which a person restricts and avoids food intake by volume and variety. The avoidance may be based on sensory attributes of food like taste, smell, texture, appearance or lack of interest in food, or past negative experience of a particular food which may lead to significant weight loss, or any nutritional deficiency or other negative health outcome.

8.3.8 Unspecified Feeding or Eating Disorders (UFED)

It includes the feeding or eating disorders that do not fit into any other category of eating disorders. However, this feeding or eating disorder may affect social, occupational, or any other area of functioning and cause distress.

8.4 WORLDWIDE STATISTICS ON FEEDING AND EATING DISORDERS

Eating disorders have received little attention, despite being the reasons for mortality and severe disability that impact not only the individual who suffers but also their caregivers. Eating disorders contribute to the years lived with disability, decrease quality of life, and increase the economic cost for individuals and their caregivers (van Hoeken and Hoek 2020). Therefore, the burden of eating disorders can be understood in terms of disability as well as mortality rates. Despite the serious health outcomes, eating disorders did not receive much attention until recently and were only included in the Global Burden of Diseases (GBD) study in 2010. According to GBD 2013, AN and BN were responsible for 1.9 million disability adjusted life years (DALYs), which is equivalent to around 1.5 million females or 41.1 per 100,000 and 0.34 million males or 8.8 per 100,000. Young females living in high-income countries showed a higher burden of eating disorders (Erskine et al. 2016). In 2017, 3.35 million healthy life years were lost to eating disorder-related disability worldwide (van Hoeken and Hoek 2020).

In GBD 2019, 13.6 million cases of AN and BN, equivalent to 176.2 per 100,000, were reported. The eating disorders accounted for 2.9 million DALYs, equivalent to 37.6 per 100,000 population (Santomauro et al. 2021).

An upward trend in cases of eating disorders has been observed. Van Hoeken and Hoek (2020) studied the decadal trend from 2007 to 2017 and found an increase of 9.4% in eating disorders. An increase of 6.2% for AN and 10.3% for BN from 2007 to 2017 was reported. Similarly, Wu et al. (2020) studied the trend in the global age-standardised prevalence rate of eating disorders and observed an increase from 172.53 to 203.20 per 100,000 population from 1990 to 2017. The age-standardised DALYs rate of eating disorders increased from 35.75 in 1990 to 43.36 in 2017 per 100,000 people, accounting for an average annual increase of 0.66.

However, these data are not a true representation of actual cases of eating disorders. Since only AN and BN were included as eating disorders in the GBD, injuries and risk factor study, therefore the data are an underestimation of the factual prevalence of eating disorders. Santomauro et al. (2021) estimated 41.9 million global eating disorder cases unrepresented in GBD 2019. BED and OSFED accounted for 17.3 and 24.6 million cases, respectively, amounting to 3.7 million DALYs globally which were ignored in GBD 2019. When these figures were added to GBD 2019 estimates, then 55.5 million people equivalent to 717.3 per 100,000 were estimated in 2019. The combined figures estimated 6.6 million DALYs, equivalent to 85.9 per 100,000 population. The burden was higher for females than males i.e., 4.7 million against 2.0 million and the burden peaked at 25–29 years for females and 30–34 years for males.

Van Eden et al. (2021), in their review article, have concluded that the incidence of AN was higher in younger people with 4% of females and 0.3% of males suffering from AN during their lifetime. Similarly, 3% of females and 1% of males suffered from BN during their lifetime. Unfortunately, the mortality rates for AN and BN are increasing. The stereotypical speculation regarding the prevalence of eating disorders among young females from Western countries has changed, as both males and females from Western as well as non-Western countries in all age groups are at risk. It has been reported that AN is common among 15–19-year-old females with high mortality risk, and BED is common in male and older individuals (Smink et al. 2012), though the severity may vary because of a couple of reasons. The Western region had 8.5 times more prevalence than the Asian region, with AN, BN, and BED 21, 7.3, and 2 times higher, respectively, in the Western region compared to the Asian region (Qian et al. 2022).

In a review article Qian et al. (2022) estimated pooled lifetime and 12-month prevalence rates of eating disorders as 0.9% and 0.43%, respectively. The lifetime prevalence of AN was 0.16% BN was 0.63%, and BED was 1.53%. They observed a fluctuating trend in the prevalence of eating disorders. An increase in the lifetime prevalence rate of EDs was reported from the 1990s (0.9%) to 2000 (2.0%), and then a decrease (0.71%) in 2010. Lifetime prevalence rates of AN increased from 0.06% in 1990 to 0.42% in 2010. Likewise, BN showed an upward trend from 0.53% in the 1990s to 1.08% in the 2010s. Nevertheless, the lifetime BED prevalence rate decreased from 1.58% in 2000 to 1.48% in 2010. Galmiche et al. (2019) concluded that the highest prevalence was of OSFED, followed by BED, BN, and AN was concluded by Galmiche et al. (2019). They reported an increase in weighted means of point prevalence for eating disorders from 3.5% in 2000–2006 to 7.8% in 2013–2018. An increase in the prevalence of eating disorders calls for recognising eating disorders as a public health concern, and understanding the risk factors associated with eating disorders so that steps for treatment and prevention could be taken up in a timely manner.

8.5   RISK FACTORS FOR FEEDING AND EATING DISORDERS

Research has shown the interplay of genetics and environment in manifesting eating disorders. Epidemiological evidence has shown that female gender, adolescents, and the affluent class are at higher risk. For a better understanding, we have classified the risks of eating disorders into five categories (Figure 8.2).

8.5.1   GENETIC FACTORS

It has been concluded through twin studies and genome-wide data that eating disorders like AN, BN, and BED run in families (Bulik et al. 2019). Genetic factors are responsible for more than a 50% chance of developing eating disorders like AN, BN, and BED (Berrettini 2004; Trace et al. 2013). The genetic and epigenetic basis for the development of eating disorders has been understood and proven through various heritability studies, linkage studies, candidate gene approach,

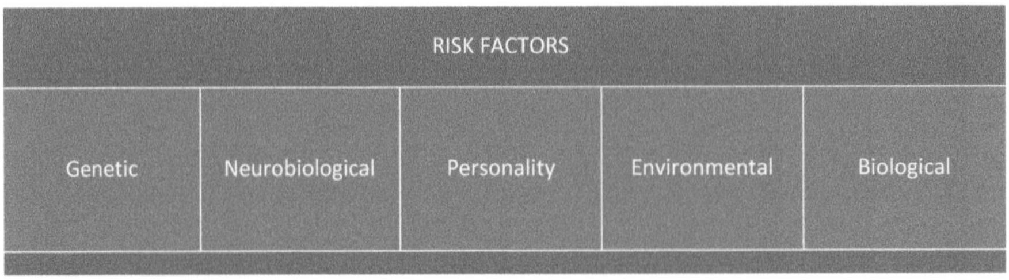

FIGURE 8.2   Risk factors for feeding and eating disorders.

genome-wide association studies (GWAS), next-generation sequencing, candidate gene, and whole-genome methylation studies (Mayhew et al. 2018).

8.5.2 NEUROBIOLOGICAL FACTORS

Neural pathways disturbing appetite regulation influence eating behaviour. According to conclusions based on neuroimaging studies (Ely et al. 2016), dysfunctions in brain circuits underlying reward and inhibition result in eating disorders. Various neurotransmitters and hormones are associated with eating disorders. Serotonin, the chemical messenger primarily known to regulate mood, also helps in appetite regulation. Serotonin imbalance can lead to eating disorders by affecting personality traits in patients, and vice versa. Restricted eating, along with excessive exercise, may alter dopamine and serotonin levels (Rikani et al. 2013).

Hormones such as sex hormones and gut hormones also affect brain response. Alteration in leptin or ghrelin influences dopamine response, which affects the approach to food in eating disorders (Frank et al. 2019). Dopamine is related to feeding behaviour (Yilmaz et al. 2015). A decrease in the reward value of food and abnormal dopamine functioning are seen in AN. The reinforcing value of food is decreased in AN, and binge eating control systems are lost in eating disorders. It has been suggested that functional and structural alterations in control circuits occur in BN purging (Steinglass et al. 2019). Food restriction, binge eating, and purging alter brain structure, affecting perception of taste, reward valuation, and energy homeostasis regulation, thereby interfering with the drive to eat. For instance, in anorexic subjects, the reward circuit is overly sensitive, and the dopamine levels trigger anxiety rather than pleasure in food intake (Frank et al. 2019).

8.5.3 PERSONALITY

Personality traits structured from both the genetic makeup as well as the experience gained from the environment have a crucial role in the development of eating disorders. Research shows that there is a difference in the personalities of subjects experiencing eating disorders compared to their healthy counterparts (Dufresne et al. 2020). These may be aetiological factors to eating disorders or may be sequelae of eating disorders (Bachner-Melman et al. 2006).

Levallius et al. (2015) compared personality traits between adult young healthy females and their disordered eating counterparts. The subjects showing eating disorders (BN and EDNOS) scored high on neuroticism and exhibited significantly high anxiety, hostility, depression, self-consciousness, impulsiveness and vulnerability compared to controls. They exhibited poor extraversion characteristics like warmth, gregariousness and positive emotions. The subjects with eating disorders were less open to the experience of fantasy, aesthetics, actions of choosing novelty, curiosity in new thoughts and ideas but were more open to willingness to re-evaluate norms and values. They showed less trust and compliance but more modesty. They scored less on conscientiousness characteristics like self-discipline, competence, dutifulness and aspiration levels compared to their healthy counterparts. In contrast, in males openness was associated with purging behaviour and emotional stability had a positive association with AN symptoms and global eating pathology. No significant association was observed between eating pathology and personality traits like conscientiousness and agreeableness in the case of males (Dubovi et al. 2016). Ham et al. (2021) concluded that personality traits are determinants of eating disorders, especially BN. Girls with a higher tendency to personality traits like perfectionism, ineffectiveness, distrust, and body dissatisfaction were inclined towards bulimic eating disorder. Anderluh et al. (2003) inferred that obsessive-compulsive personality trait suffered during childhood is an important risk factor in the development of eating disorders, especially AN in later life. Likewise, confirmation was made by Bachner-Melman et al. (2006) who found that in women aged 14–36 suffering from AN or had a past history of AN showed perfectionism, harm avoidance, fear of failure, obsessive and low self-esteem traits compared to control. These personality factors were potential risk factors in the development of eating disorders.

Perfectionists and people in whom the feeding pattern is controlled by their moods are more prone to the risk of developing eating disorders (Upadhyah et al. 2014).

Mas et al. (2011) structured a model showing the relationship between personality traits and eating disorders. Individuals exhibiting schizoid, self-destructiveness, borderline personality traits, and paranoid behaviour had low self-esteem, and borderline personality traits lead to perfectionism and body dissatisfaction. These three factors, i.e., perfectionism, body dissatisfaction, and low self-esteem further resulted in dietary restraint and purging behaviour. Subjects suffering from AN had greater exposure to perfectionism, and those suffering from BED had conduct problems compared to normal healthy counterparts (Hilbert et al. 2014).

8.5.4 ENVIRONMENTAL FACTORS

Industrialisation and urbanisation fuelled by socioeconomic growth, exposure to media idealising thin body frames especially for females, and the influence of the fashion and beauty industries have led to societal transformation in Asian countries, thereby augmenting eating disorders (Pike and Dunne 2015). Eating disorders, once thought to be a disorder of the Western world have sneaked into less affluent countries because of rapid globalisation.

Sociocultural factors also play an important role in the development of eating disorders. In some cultures, female beauty is idealised with extreme thinness. Individuals exposed to such idealisation at any age and through any environmental source followed by internalisation of the ideal beauty body frame, comparing and fitting oneself in that frame by practising extreme behaviours like dietary restriction, purging, extreme exercise, or use of medication, etc., is the course of developing eating disorders. Failure in an attempt to attain the ideal beauty frame leads to body dissatisfaction, strict dietary restriction, or overeating followed by dietary restraint and purging. During this process, guilt, anxiety, and low self-esteem set in the individual and the interplay of psychological mediators further aggravates the condition. Further exposure to an environment like a family promoting a thin body frame as ideal, or unresolved family conflicts, school environments where either sometimes teachers emphasise thin body frames especially for the students who are slightly overweight or peer teasing, exposure to social media endorsing standards in the beauty and fashion industry, abuse especially sexual abuse, any trauma, pressure to be thin while working in professions like the fashion industry, athletics, or the armed forces, inability to cope with the stressors are responsible for the development of eating disorders (Striegel-Moore and Bulik 2007; Mayhew et al. 2018).

Adoption of socioculturally promoted methods like fasting for food avoidance were seen as a risk factors among adolescent girls in India. Religious fasting every week or for more than one day in a week was the best method for the school-going girls to avoid food without being noticed by the family. Moreover, it was promoted by society as it was seen in conformity with social norms, especially in rural and less-developed areas (Pathak et al. 2020).

Children of mothers with eating disorders are also at risk of developing eating disorders. The mother's eating disorder has an impact on the child's eating behaviour. Such children experience socioemotional difficulties and are at a higher risk of psychopathological problems (Martini et al. 2020).

Parental attitude also influences the occurrence and severity of eating disorders in children. Parental over-concern regarding eating behaviours and parental perception of the child being overweight lead to undereating or early feeding problems in children. Parents who themselves have weight and eating-related attitudes and behaviours increase the risk of developing eating disorders in children (Herzog and Eddy 2009). Overprotective, intrusive, controlling, emotionally unresponsive parents are risk factors for the development of eating disorders in children.

Individuals exposed to the pressure to be thin coming from peers, media, and parents are at higher risk (Upadhyah et al. 2014). Parental overprotection and lack of care increases the vulnerability towards anorexia in school-going girls (Pathak et al. 2019). Parental criticism has a great impact. High parental expectations, minimal affection from parents, under-involved parents, separation from parents, low maternal care, and high overprotection have been associated with BED. Abuse (physical, substance, and sexual), bullying, and criticism about appearance, shape, weight, and eating can lead to development of eating disorders like BED and BN. In summary, it can be said that any type of early-life traumatic or stressful events are risk factors for the development of eating disorders (Fairburn et al. 1998; Hilbert et al. 2014; Solmi et al. 2021).

Recent studies show that socioeconomic status, once thought to be an important risk factor in the development of eating disorders, is no longer an important determinant. Eating disorders are universally experienced irrespective of socioeconomic status (Mulder-Jones et al. 2017). AN was known as a disease of affluence, but no such association was found between socioeconomic status and eating disorders. (Striegel- Moore and Bulik 2007). On the other hand, Ham et al. (2021) found that low socioeconomic status is predictive of BN behaviour among girls.

8.5.5 BIOLOGICAL FACTORS

Certain biological factors like age, gender, puberty, and race predispose to the risk of developing eating disorders.

Adolescents and young adults are more prone to eating disorders; however, the chances of developing a disorder in later age are also possible. It was seen that with the increase in age, thin ideal internalisation, dieting, and psychosocial functioning decrease, but body dissatisfaction increases with age (Rohde et al. 2017). AN and BN occur during adolescence, and BED in adulthood (Striegel-Moore and Bulik 2007). The onset age for AN was found to be lower (18.9 years), followed by BN (19.7 years) and BED (25.4 years) (Hudson et al. 2007). BN behaviour was seen as early as 11–12 years of age among girls (Ham et al. 2021). Early age at menarche was seen as risk factor for developing ED (Upadhyah et al. 2014). In a simulation study, it was seen that the highest estimated mean annual prevalence of eating disorders occurred at 21 years of age, and the mean lifetime prevalence increased by age 40, indicating that eating disorders are at their peak in the early 20s and then decrease with advancement in age (Ward et al. 2019).

Gender influences the propensity to the development of eating disorders. According to Ward et al. (2019), females experience more episodes of eating disorders compared to males, with AN and BN more prevalent in males and BED more prevalent in females. One in seven males and one in five females experiences eating disorders by the age of 40. Some 90% of those afflicted with AN are females. Males also exhibit eating disorder symptomology; however, it remains ignored: 2.1% of college men aged 18–26 years exhibited eating disorder pathology with purging behaviour, 9% showed compensatory fasting, and 14.9% engaged in excessive exercise (Dubovi et al. 2016). However, the characteristics of eating behaviour differ in men and women. Body checking, body avoidance, fasting, binge eating, and vomiting to avoid weight gain in weight after binge eating were mainly observed in women, whereas men exhibited higher overeating. The use of laxatives and exercising was the same for both genders (Striegel-Moore et al. 2009). Studies have shown that females exceed males in experiencing eating disorders, especially for AN and BN where weight control is important. However, no such gender difference was seen for BED (Striegel-Moore and Bulik 2007).

Exploring the role of race as a risk factor in the development of eating disorders, it was seen that being Hispanic acts as a moderator for thin body preoccupation and social pressure, which consequently leads to development of eating disorders (McKnight Investigators 2003). BN was found to be more prevalent in African-American girls rather than white girls (Ham et al. 2021).

The common perception that eating disorders are prevalent at a particular age, gender, and race has been challenged by recent research. Mixed results have been found, and the eating disorders cannot now be restricted to specific demographics. They can affect anyone, irrespective of age, gender, and race.

None of the above-mentioned aetiological factors can be studied in isolation in development of eating disorders. An interaction among these factors leads to a predisposition to eating disorders. Mayhew et al. (2018) created a framework suggesting the interplay of genetic and environmental conditions under which eating disorders will occur. According to them, genes predisposing to AN in the presence of the environmental condition of pressure to be thin, will lead to AN (restrictive subtype), whereas in absence of pressure to be thin, no disease may develop. Genes predisposing to binge eating, coupled with the condition of pressure to be thin, will lead to BN, and in the absence of pressure to be thin, will lead to BED. Genes predisposing to both AN and binge eating, along with the pressure to be thin, will lead to AN (binge/purge subtype), and under the absence of pressure to be thin, will lead to BED. A thinness-promoting environment may lead to an overexpression of genes that suppress appetite in those who are already genetically susceptible to weight suppression (Yilmaz et al. 2015).

8.6   TREATMENT AND PREVENTION

Seeing the multifaceted aetiology of eating disorders, the treatment and prevention call for the shared responsibility of a multidisciplinary team comprising mental health clinicians, dietitians/nutritionists, and general medical clinicians. The treatment therapies for feeding and eating disorders can be broadly classified as shown in Figure 8.3.

8.6.1   PHARMACOTHERAPY

Researchers have seen the potency of certain drugs in treating eating disorders. The psychological comorbidities existing as aetiology to or outcomes of eating disorders need to be treated with

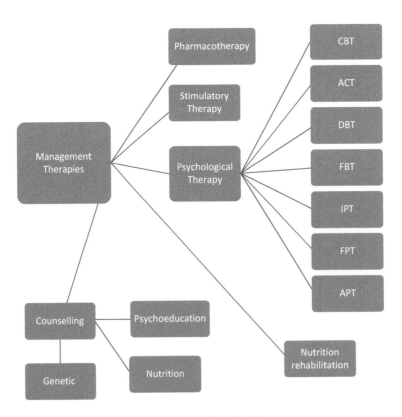

FIGURE 8.3   Management therapies in feeding and eating disorders.

psychotropic medication. Clinical research has proven the efficacy of antidepressant medications like fluoxetine, selective serotonin uptake inhibitors (SSRIs), tricyclic antidepressants (TCAs), monoamine oxidase inhibitors in reducing the binge–purge episodes in BN. Antiepileptic medications like topiramate in combination with phentermine show improvement in BN and BED. Antidepressants, weight management drugs, and stimulant medication are used in treating BED, as they are seen to reduce binge eating and suppress appetite. Opioid antagonists like naltrexone have the potential to decrease body weight by reducing calorie intake in BN and BED. Second-generation antipsychotics like olanzapine showed an impact on weight gain and other psychiatric disorders associated with AN (Davis and Attia 2017; Lutter 2017; Milano et al. 2013). Olanzapine is related to changes in the dopaminergic and serotonergic system in subjects with AN (Copur and Copur 2020). However, pharmacotherapy does not show promising results in AN as in the case of BN and BED (Gorla and Mathews 2005).

Owing to the relapsing nature of eating disorders and safety concerns associated with drug intake, focus on other potential therapies should be given.

8.6.2 STIMULATORY THERAPIES

Beside pharmacotherapy, certain non-invasive techniques like transcranial direct current stimulation, transcranial magnetic stimulation, and deep brain stimulation are also considered as a means of treating eating disorders. These are neuromodulatory techniques with the potential to treat eating disorders and their associated psychological comorbidities. Hormones responsible for increasing hunger or suppressing appetite may also have a beneficial role in treating eating disorders. Administration of ghrelin (hunger hormone) and oxytocin (hypothalamic neuropeptide) may improve calorie intake, food consumption, and thereby body weight in AN. Similarly, agonists of glucagon like peptide-1 (gut hormone and polypeptide YY) are seen to increase satiety and suppress appetite and can be used as a treatment in BN and BED (Lutter 2017). However, this research is still in its infancy and more clinical trials are required before establishing their efficacy.

8.6.3 PSYCHOTHERAPIES

8.6.3.1 Cognitive Behaviour Therapy (CBT)

CBT is a method of psychosocial intervention. CBT is based on the principle that the root cause of psychological problems is based on faulty ways of thinking and behaving, and therefore making changes in thinking and behaviour patterns helps in coping with the disorders. In CBT, the therapist and the patient work as a team to bring about changes in thinking and behaviour of the patient and to cope effectively with problematic emotions (APA 2017). With time and advancements in the understanding of eating disorders and their challenges, the transdiagnostic perspective was proposed, and enhanced cognitive behaviour therapy (CBT-E) was seen as a treatment for all forms of eating disorders (Fairburn et al. 2003).

CBT-E is a more effective enhanced version of CBT. It is based on the transdiagnostic theory designed to treat all eating disorders. It is more effective in terms of the usage of variety of new procedures to improve outcome and also includes modules to address external obstacles to change, like perfectionism and low self-esteem. There are two forms of CBT-E. First, the focused form (CBT-Ef) which addresses the eating disorder pathology, and second, the broad form (CBT-Eb) which addresses external obstacles to change along with the core eating disorder pathology. CBT-E aims to change the behaviour of individual to modify the thinking and behaviour pattern. It is a four-stage treatment in which the patient is engaged in the treatment. Emphasis is given to educating the patient regarding the disorder and its treatment, and self-monitoring (weekly weighing and regular eating) in the first stage. In the brief second stage, problems and barriers to change are identified and changes in process are made if necessary. In the third stage, the focus is on the processes maintaining eating disorders. Problems like over-evaluation of shape and weight and their origin are addressed. Rigid dietary rules and their triggers like outside events or changes in mood are treated. External processes like perfectionism, low self-esteem, and interpersonal problems that may maintain eating disorders are tackled. In the last stage, efforts are made to maintain the progress made by the patient and trim down the risk of relapse (Murphy et al. 2010). The effectiveness of CBT-E has been proven in multiple studies (Fairburn et al. 2015; de Jong et al. 2018; Frostad et al. 2018).

8.6.3.2   Acceptance and Commitment Therapy (ACT)

It is a form of psychotherapy based on the principle of accepting thoughts and feelings rather than running away or feeling guilty for them. ACT is based on the concept that mental disorders occur due to aversive internal experiences and therefore increasing one's psychological flexibility in being aware and accepting the situation and performing mindfulness strategies to commit and make behavioural changes. This is the key to counteracting the problem. ACT helps individuals to commit actions in accordance with their values. ACT is an effective intervention strategy in eating disorders. In eating disorders, the patient's behaviour is aimed at avoiding or controlling inner experiences by either restricting food intake or practising binge–purge behaviour. ACT helps the patients to change behaviour aimed at controlling or avoiding unwanted and intrusive thoughts and emotions. It helps the patients to accept them and choose behaviour in accordance with their values. Dealing with overevaluation of shape and weight and their control through diet needs acceptance and motivation on the part of the patient. Mindful acceptance counteracts experiential avoidance by increasing psychological flexibility (Fogelkvist et al. 2020; Ackerman 2017; Shumlich 2017). The effectiveness of ACT in eating disorders has been seen by researchers as this technique aims to decrease avoidance and increase awareness and motivation (Juarascio et al. 2010; Juarascio et al. 2013; Masuda et al. 2014).

8.6.3.3   Dialectical Behaviour Therapy (DBT)

Mood intolerance is recognised as the key factor in the onset of eating disorders, and therefore DBT intervention deals with emotional dysregulation. Dietary restraints or binge eating and their compensatory behaviour are outcomes of mood fluctuation. Diet becomes a coping strategy for emotional dysregulation. DBT focuses on experiencing and controlling unbearable emotions in a healthy way. The therapy aids in inhibiting inappropriate behaviour like restricted eating or binge eating in response to negative emotions and increases conscious control (Shumlich 2017). A significant decrease in binge–purge behaviour was seen after dialectical behaviour therapy (Safer et al. 2001). DBT was proven to be effective in women with borderline personality disorder with AN ands BN (Kroger et al. 2010) and BED (Rahmani et al. 2018). It helps in emotion and BMI regulation and is seen to be less time-consuming and cost-effective compared to other intervention techniques like CBT (Rahmani et al. 2018; Lammers et al. 2020).

8.6.3.4   Family-Based Treatment (FBT)

FBT is also known as the Maudsley method of treatment and is found to be effective in treating eating disorders, especially AN in adolescents. The treatment consists of three stages. In the first stage, the aim is the restoration of healthy eating and weight in children. During this stage, parents play a crucial role in helping the children provide food of appropriate quality and quantity. The parents bear the responsibility of monitoring the timing and frequency of meal consumption by the child. In the second stage, the responsibility of consuming food is given back to the child under the parents' supervision. In the last and third stage, normal family life is resumed. However, the parents are required to help the adolescent in identifying and meeting the challenges of development. The important part is to ensure that eating disorder relapse does not occur and the child does not adopt an eating disorder as a coping strategy for the challenges of life. FBT is equally beneficial in BN and AFRID where the focus is on reducing the binge–purge pattern and increasing the variety of food, respectively. In each type of disorder, the responsibilities lie with the family along with the therapist to normalise the eating pattern. Unlike other described psychotherapies, FBT is based on the principle of what needs to be done for the recovery rather than addressing the cause (Loeb and le Grange 2009; Rienecke 2017).

8.6.3.5 Interpersonal Psychotherapy (IPT)

This therapy deals with the analysis of eating disorders in the context of interpersonal difficulties. It is based on the principle that interpersonal and social issues lead to eating disorders and therefore strategies to improve interpersonal skills will help to overcome eating disorders. Rieger et al. (2010) proposed the model of IPT for eating disorders. According to them, negative self-evaluation and poor self-evaluation which develop due to poor interpersonal skills trigger eating disorders. Eating disorders act as a maladaptive mechanism to overcome negative self rather than through adaptive interaction with the social world. Poor interpersonal skills are associated with an increase in body image concerns leading to the development of eating disorders which further may lead to a negative self. Thus, negative self-evaluation can be a cause as well as a consequence of eating disorders. IPT focuses on understanding the interrelationship between interpersonal deficit and eating disorders. IPT helps the patient improve their interpersonal life which in turn increases their self-esteem and self-valuation, thereby decreasing the eating disorders (Murphy et al. 2012). IPT was found to be effective in BED (Wilson et al. 2010), feasible in preventing excess weight gain (Tanofsky-Kraff et al. 2010) and is a cost-effective alternative treatment with long-term stable efficacy (Miniati et al. 2018).

8.6.3.6 Focal Psychodynamic Therapy (FPT)

FPT is based on the theory that the psychological disorders in the present are related to unresolved conflicts in the past. FPT addresses unconscious or preconscious conflicts that trigger the disorder. Repressed or defended motives, emotions, and feelings encountered during early life account for psychological disorder in later life, and hence disordered eating becomes a way to cope with them.

During the therapy, the focus is found by exploring the patient's response to recent stress, which tells us about the patient's habitual character and mode of defence and adaptation. This helps the therapist to understand the patient's character structure and focus on the area of the patient's problem that has led to the present disorder (Chernus 1983; www.ucl.ac.uk). The treatment is divided into three phases. The first deals with building a therapeutic alliance, the ego-syntonic nature of the disorder, and self-esteem. The second phase focuses on the association between interpersonal relationships and eating behaviour, and third with the termination of treatment and the beginning of normal life (Wild et al. 2009). Zipfel et al. (2014) found that focal psychodynamic therapy was useful in increasing weight and BMI in anorexic subjects along with its effectiveness in recovery during follow-up.

8.6.3.7 Adolescent and Parent Treatment (APT)

APT is a novel treatment option still in its nascent stage. It addresses the physical and psychological recovery of adolescents suffering from AN. It focuses firstly on intensive parental refeeding with the help of a dietitian and therapist and secondly on the psychological, emotional, temperamental, and developmental needs of adolescents. APT is a 20–30 session treatment programme continued for a period of 9–12 months (Ganci et al. 2021).

Various psychological interventions are being intensively studied for their efficacy in the treatment of eating disorders. Some of the therapies are well established and few are still in their budding stage with promising potentiality. Depending upon the type of eating disorders, the type of therapy may also change. In a systematic review (Costa and Melnik 2016) it was found that Maudsley Family Therapy was most effective in AN. Besides focal psychoanalysis, supportive psychotherapy, and CBT were also helpful in AN. CBT was observed to be the most effective treatment in BN and BED. IPT, DBT, supportive psychotherapy, and self-help manuals were also beneficial in eating disorders like BN and BED. CBT, IPT, and DBT are effective in treating BED (Marzilli et al. 2018). Since there are multiple effective methods of treating eating disorders, Jansingh et al. (2020) placed emphasis on following the shared decision making (SDM) protocol for selecting the best method involving patients actively increasing self-motivation. Thus, it can be concluded that the selection and effectiveness of intervention depends on the severity of the problem and associated psychosocial and biological comorbidities with a joint effort from the therapist and patient.

8.6.4 Nutritional Rehabilitation

Nutritional rehabilitation is very important in eating disorders. The faulty eating habits in eating disorders adversely affect the nutrient status. Restoration of body weight and replenishment of all the nutrients are the major goals of nutritional rehabilitation.

The core component of nutritional rehabilitation in AN is to promote metabolic recovery, restore body weight, maintain the gained body weight, normalise the eating pattern, understand the perception of hunger and satiety, reverse the medical complications, and improve the psychological functioning. In AN, the refeeding course includes deposition of lean body mass followed by deposition of adipose tissue. Refeeding syndrome is a complication encountered, which could be managed by slow feeding along with monitoring of body weight and electrolyte balance (Golden and Meyer 2004; Marzola et al. 2013).

In a review article by Marzola et al. (2013), it was concluded that patients with AN had deficient levels of macro and micro nutrients and therefore the nutritional rehabilitation goals should be a gain in weight at the rate of 0.24–0.45 kg per week for outpatients and 0.9–1.36 kg per week for hospitalised patients as per APA guidelines. Calorie intake should be 30–40 kcal per kg per day initially advancing to 70–100 kcal per kg per day. Multivitamin-mineral supplements should be given. The importance should be given to adding variety to the diet along with maintaining nutritional adequacy. Furthermore, to maintain weight, 110–140 g carbohydrate, 15–20 g essential fatty acids, and 1 g protein per kg body weight daily are recommended. High biological value protein, omega-3 and omega-6 fatty acid-rich food, and variety of carbohydrates like complex carbohydrates as well as simple carbohydrates are suggested. Fluid balance needs to be maintained. Regular monitoring to avoid any clinical complications and vitamin toxicity is needed. Supplementation of vitamin D, magnesium, and dietary fibre is also very essential to maintain the microbiota, which becomes altered in eating disorders like AN (Frank et al. 2019).

In AN, repeated cycles of starvation, in BN recurrent binge and purge cycles, and in BED overeating alter the normal physiological functioning. The depletion or excessive accumulation of nutrients may affect the functioning of the heart, kidneys, digestive system, and lead to various secondary health issues like hypotaemia, hypoglycaemia, delayed puberty, delayed and irregular menses, diabetes, peptic ulcer, hypertension, dyslipidaemia, peptic ulcer, gall bladder disease, constipation, tooth decay, osteoporosis, etc. (Rome and Ammerman 2003). Thus, nutrition care can help in managing the primary and secondary health issues of eating disorders.

Nutrition rehabilitation is a slow process and depends on the severity of the anorexic state. In the case of severe anorexia, the need is to hospitalise the patient and give intravenous feed to restore fluid and electrolytes and prevent dehydration. Peripheral parenteral nutrition is suggested to support oral intake. The nutrition care should start with a liquid diet and proceed towards a soft to a full diet. The aim is to develop normal eating patterns without putting any psychological pressure on the patient (Mudambi and Rajagopal 2009; Mehler et al. 2010). For proper refeeding of anorexics and to avoid refeeding syndrome, dietitians must make sure that calorie intake should rarely exceed 70–80 kcal/kg body weight and protein intake should not exceed 1.5–1.7 g per kg body weight and in total parenteral nutrition carbohydrate intake should not exceed 7 mg/kg/min. The diet should be low in sodium and judicious use of fibre in the diet should be promoted (Mehler et al. 2010).

Dietitians should also bring bulimics into a normal eating pattern and behaviour. Maintenance of a normal body weight and restoration of all essential nutrients is targeted. The basic food guide should be followed. Meal intake in bulk should be avoided; instead, personalised meal patterns, i.e., three meals and three snacks, should be followed. Importance should be given to adding variety to the diet, keeping in mind the likes and dislikes of the patients (Mudambi and Rajagopal 2009). Nutrition intervention for BN and BED is almost similar. Regular eating based on a personalised meal plan of three main meals with two or three snacks helps to disguise the long gap between meals and lowers the swings between hunger and satiety, thereby decreasing binge–purge episodes. In AFRID, besides restoring weight (if weight loss exists), the major nutritional therapy is to identify the foods avoided by patients and gradually reintroduce them. The new or avoided foods are added to the meal as well as snacks after they are completely explored by the patient (Academy for Eating Disorders Nutrition Working Group nd).

Certain food plans have been designed for eating disorders with the aim to meet energy and nutrient requirements, provide an organised approach to food consumption, and accustomisation with the feared, binged and purged foods. These food plans are exchange-based systems where food is organised into meals and snacks; rule of three system where normal meals are tailored to meet nutrient requirements and snacks like desserts to add pleasure in eating; plate-by-approach wherein parents are instructed to fill in a ten-inch plate with prescribed proportions of food like 50% grains, 25% protein, 25% fruit and vegetables with fat and dairy for AN or AFRID child. Similarly, for BN and BED subjects, the plate consists of equal proportions of grains (1/3), protein (1/3), and fruit and vegetables (1/3) (Academy for Eating Disorders Nutrition Working Group nd). Nutritional care and appropriate diet are integral parts of eating disorder treatments, and therefore dietitians along with medical professionals play a crucial role in managing the ill effects of eating disorders and bringing the patients into the mainstream.

8.6.5 Genetic Counselling

Genetic counselling should be imparted to make the individual understand the genetic makeup related to eating disorders, the associated risk factors, and ways to reduce the risk which may, in turn, assist in preventing the development of eating disorders and aid recovery in individuals suffering from an eating disorder. Genetic counselling is a process of making people aware of the familial implications of genetic contribution to disease. Genetic counsellors focus on addressing the guilt, shame, stigma, blame, or fear associated with the cause of the condition in the patient's family and help patients in making the best decisions. Genetic counselling reduces the feeling of stigma, shame, guilt, and encourages the patient to seek help and in fast recovery (Bulik et al. 2019; Michael et al. 2020).

8.6.6 Psychoeducation

Psychoeducation is a process of imparting education and information about a medical or psychological condition to individuals and their family members. It is very essential in managing disorders and can play a vital role in treatment as well as prevention. The basic aim of psychoeducation is to inform the patients and their family members and caregivers about the illness, its origin, and its treatment process so that the concerned person can take charge of the treatment process rather than considering it as a shameful condition. Psychoeducation helps to decrease the rate of hospitalisation, the rate of relapse and increases compliance with the treatment plan. It enables the subjects to understand the problem and make behavioural changes with the support of family (Bauml et al. 2006; Shindhe et al. 2014). Psychoeducation was found to be effective in various eating disorders. Scientifically driven, patient-centred psychoeducation helped patients suffering from eating disorders (Belak et al. 2017). It improved body dissatisfaction, bulimic response, and affect regulation (Cruchet et al. 2021; Storch et al. 2011).

Since the family plays an important supportive role in managing eating disorders, especially in children and adolescents, psychoeducation for family members also becomes crucial. It lowers distress, improves interaction within the family, and helps in dealing with societal pressure. Parents and caregivers get equipped with the skills to manage stress and emotion dysregulations (Uehara et al. 2001; Spettigue et al. 2015).

8.6.7 Nutrition Counselling

Apart from the nutritional assessment, providing nutrition care and monitoring, nutritionists have a major role in nutrition counselling of patients suffering from eating disorders. Nutrition counselling helps in establishing a healthy eating pattern and restoration of psychophysiological wellbeing. Nutrition counselling not only encourages regular eating but also discourages dieting. It makes the patients aware of the ill effects of starvation, dieting, binging–purging, medical risks associated with disordered eating, and other comorbidities. The process is done gradually without putting any stress on the patient, and therefore nutrition education can play a very significant role in treating, managing, and preventing eating disorders.

8.6.8 Adjunctive Therapies

Complementary and alternative therapies can be used as adjunctive therapies in appeasing the symptoms secondary to eating disorders. Therapies like yoga, acupuncture, hypnosis, and spiritual healing can be used as adjunctive therapy either to address core symptomology of eating disorders or associated comorbidities (Madden et al. 2014). Few studies have been done in this arena. Massage and bright light therapy can be used to alleviate depression in BN. Acupuncture and relaxation therapy in treating state anxiety in individuals with AN (Fogarty et al. 2016). Yoga may also provide benefits in eating disorders, especially AN (Rizzuto et al. 2021). Yoga helps in relieving stress, depression, and creates body awareness. Certain herbs which may have a role in regulating hunger and satiety can be explored. Cinnamon, vinegar, and oxymol are said to act as appetisers and should be taken before meals by anorexics. Herbs of warm temperament may help the anorexic patients (with cold temperament) by increasing their appetite (Nimrouzi and Zarshenas 2018). Research linking adjunctive therapies with eating disorders is quite scarce and further studies are warranted to prove their efficacy in treating and managing eating disorders.

8.7 SUGGESTIONS TO MANAGE AND PREVENT EATING DISORDERS

8.7.1 Identification of the Risk Group

Early identification of individuals at risk and early intervention will help in tackling eating disorders. For instance, children of mothers with eating disorders are at a higher risk of developing an eating disorder and thus, by identifying those children, prevention can be planned (Martini et al. 2020). Early detection of subjects (children, adolescents) and initiating primary prevention strategies are suggested, especially in BED, where a broad range of adverse effects such as obesity, social impairment, depression, anxiety, emotional distress, substance abuse, self-harm, and suicidal tendency is high (Marzilli et al. 2018).

8.7.2 Adoption of Successful Prevention Programmes

An understanding of modifiable risk factors can form the cornerstone of managing and preventing eating disorders. Ciao et al. (2014) have revealed nine prevention programmes for eating disorders that were successful in reducing and preventing eating disorder pathology. The prevention programmes were based on social cognitive or cognitive behavioural or cognitive dissonance theory, focusing on reducing risk factors and making positive behavioural changes. The successful prevention programmes were The Weigh to Eat, Stewart's untitled programme, Planet Health, Student Bodies, The Body Project/Dissonance Programs, Healthy Weight, New Moves, Yager's untitled programme, and Eating, Aesthetic Feminine Models, and the Media. The content of these successful programmes can be further exploited and modified to form the base of prevention programmes in respect to the target population (age, gender), and other associated modifiable risk factors.

8.7.3 Designing Prevention Programmes

Prevention programmes designed to strengthen social skills, healthy self-perception, and self-esteem among children are required. Prevention programmes should be aimed at building a positive body image and accepting bodily changes. The prevention programme must include cognitive interventions to alter maladaptive attitudes and dysfunctional behaviour among children. Parents, family members, and teachers should be involved in the prevention programmes. Neugebauer et al. (2011) have proposed a prevention model for eating disorders. The model is suggested for elementary school children, involving their parents and teachers in four sessions twice a year. The first three sessions, didactic and interactive in nature, are for parents and mentors, and the fourth for the children.

8.7.4 Imparting Awareness at an Early Stage

Education regarding the ill-effects of eating disorders should be imparted to children in schools. Young children need to be made aware of a healthy body and a healthy mind because studies show that eating disorders set in at an early age. Therefore, the child must be prepared for handling all the risk factors that may predispose eating disorders. The imparting of awareness has to be done very meticulously because, as it is seen, most parts of the world are undergoing a transition. Obesity and overweight are also topics of health concern, and it was seen in a study (Ham et al. 2021) that women who participated in a preventative educational programme on the ill-effects of overweight reported bulimic behaviour. Thus, awareness has to be provided conscientiously, weighing the pros and cons.

8.7.5 What Can the Government or Stakeholders Do?

Stop advertising or glamorising the slim body concept. The association of beauty with a slim body needs to be curtailed. The fashion industry, the media, and beauty products companies need to take responsibility for understanding and differentiating between a healthy body and a body with eating disorders, projecting the same which may otherwise bring pressure on youngsters to be, dissatisfied with their bodies, and therefore create psycho-physiological health issues. Certain policies should be introduced by governments to check the hype crafted by fashion industries.

8.7.6 What Can Parents, Family, and Immediate Environment Do?

Parents, the key players in shaping the personality of a child, must equip the child with the knowledge of good nutrition and bad nutrition from childhood and make them aware of their bodies. Parents can help the child by diverting their attention towards other artful and recreational activities which interest the child.

Imparting psychoeducation to parents of children suffering from eating disorders will assist parents in managing the emotional stress and improving the functional coping of parents with their child's disorder (Holtkamp et al. 2005). It will encourage the parents to seek professional help without any stigma and emotional stress. It is seen that parent group intervention for six weeks improved the skills, understanding, confidence, and knowledge regarding eating disorders among parents, and parents could effectively make the child adhere to meal plans (Nicholls and Yi 2012).

8.8 CONCLUSION

Feeding and eating disorders are one of the major health challenges today, affecting children, adolescents, and youth in particular. Eating disorders are one of the main causes for an increase in morbidity and mortality. They have adverse effects on physiological, psychological, and social functioning. Feeding and eating disorders are predisposed by the interplay of genetic and environmental factors, along with other biological and personality factors which augment the risk of developing eating disorders. A comprehensive treatment approach is required because of the relapsing nature of eating disorders. Moreover, eating disorders have a bi-directional relationship with psychological morbidities. Various psychological problems may be aetiology to or outcomes of an eating disorder. A multidisciplinary team of psychologists, medical professionals, nutritionists, and dieticians is required to manage eating disorders. Currently, a few pharmacological therapies are known to treat eating disorders. Besides, many established psychological therapies are recognised in managing the disorder, while a few effective therapies are continuously emerging. Nutrition rehabilitation is another significant component in treating and managing eating disorders. Nutritionists and dietitians play a major role in maintaining the optimum nutritional state essential for growth and development. Counselling and psycho-nutrition education are indispensable in treating and preventing the onset of eating disorders. Parents, family, caregivers, and teachers who form the immediate environment of the patient (child/adolescent) should be educated to empathise with the situation, understand the explanation of origin without associating any stigma with eating disorders. This will help the youngsters to deal with the disorder, avoid relapse, and prepare them to face the developmental challenges in healthy ways. Authors also suggest a need to explore the possibilities of adjunctive therapies like yoga, meditation, etc. in managing feeding and eating disorders. These clinically proven adjunctive therapies could serve as cost-effective strategies in managing core symptomatology along with other secondary symptoms.

Thus, it can be summarised that although we can categorise feeding and eating disorders as deadly in terms of mortality and morbidities, their occurrence can be curtailed and treatment can be accelerated by pharmacotherapies and non-pharmacotherapies. Health professionals, family, and society can jointly work in the venture of treating and preventing feeding and eating disorders.

REFERENCES

Academy for Eating Disorders Nutrition Working Group. n.d. Guidebook for nutrition treatment of eating disorders, pp. 1–59. https://higherlogicdownload.s3.amazonaws.com. (accessed 16 March 2022).

Ackerman, C.E. 2017. How does acceptance and commitment theory (ACT) work? www.positivepsychology.com (accessed 28 May 2022).

Anderluh, M.B., K. Tchanturia, S. Rabe-Hesketh, and J. Treasure. 2003. Childhood obsessive-compulsive personality traits in adult women with eating disorders: Defining a broader eating disorder phenotype. *American Journal of Psychiatry* 160, no. 2: 242–247. https://doi.org/10.1176/appi.ajp.160.2.242

APA. 2013. *Diagnostic and Statistical Manual of Mental Disorders*, 5th edition. Washington, DC: American Psychiatric Press.

APA. 2017. What is cognitive behavioral therapy? www.apa.org/ptsd-guideline/patients-and-families/cognitive-behavioral.pdf (accessed 18 March 2022).

Bachner-Melman, R., A. Zohar, I. Kremer, and R. Ebstein. 2006. Psychological profiles of women with a past or present diagnosis of psychological profiles of women with a past or present diagnosis of Anorexia Nervosa. *The Internet Journal of Mental Health* 4, no. 2: 1–9.

Bauml, J., T. Frobose, S. Kraemer, M. Rentrop, and G. Pitschel-Walz. 2006. Psychoeducation: A basic psychotherapeutic intervention for patients with schizophrenia and their families. *Schizophrenia Bulletin* 32: S1–S9. https://doi.org/10.1093/schbul/sbl017

Belak, L., T. Deliberto, M. Shear, S. Kerrigan, and E. Attia. 2017. Inviting eating disorder patients to discuss the academic literature: A model program for psychoeducation. *Journal of Eating Disorders* 5: 49. https://doi.org/10.1186/s40337-017-0178-7

Berrettini, W. 2004. The genetics of eating disorders. *Psychiatry* 1, no. 3: 18–25.

Bulik, C.N., L. Blake, and J. Austin. 2019. Genetics of eating disorders. What the clinician needs to know. *Psychiatric Clinics of North America* 42, no. 1: 59–73.

Chernus, L.A. 1983. Focal psychotherapy and self pathology: A clinical illustration. *Clinical Social Work Journal* 11, no. 3: 215–227.

Ciao, A.C., K. Loth, and D. Neumark-Sztainer. 2014. Preventing eating disorder pathology: Common and unique features of successful eating disorders prevention programs. *Current Psychiatry Reports* 16, no. 7: 453. https://doi.org/10.1007/s11920-014-0453-0

Copur, S., and M. Copur. 2020. Olanzapine in the treatment of anorexia nervosa: A systematic review. *Egyptian Journal of Neurology, Psychiatry and Neurosurgery* 56: 60. https://doi.org/10.1186/s41983-020-00195-y

Costa, M.B., and T. Melnik. 2016. Effectiveness of psychosocial interventions in eating disorders: An overview of Cochrane systematic reviews. *Einstein* 14, no. 2: 235–277. https://doi.org/10.1590/S1679-45082016RW3120

Cruchet, L., E. Scanferla, A. Laszcz, P. Gorwood, and L. Romo. 2021. Remote psychoeducation for eating disorders: An exploratory study during lockdown. *European Psychiatry* 64, no. S1: S701–S702.

Davis, H., and E. Attia. 2017. Pharmacotherapy of eating disorders. *Current Opinion in Psychiatry* 30, no. 6: 452–457. https://doi.org/10.1097/YCO.0000000000000358

de Jong, M., M. Schoorl, and H.W. Hoek. 2018. Enhanced cognitive behavioural therapy for patients with eating disorders. *Current Opinion in Psychiatry* 31, no. 6: 436–444.

Dubovi, A.S., Y. Li, and J.L. Martin. 2016. Breaking the silence: Disordered eating and big five traits in college men. *American Journal of Men's Health* 10, no. 6: NP118–NP126. https://doi.org/10.1177/1557988315590654

Dufresne, L., E.L. Bussieres, A. Bedard, N. Gingras, A. Blanchette-Sarrasin, and C. Begin. 2020. Personality traits in adolescents with eating disorder: A meta-analytic review. *International Journal of Eating Disorders* 53, no. 2: 157–173.

Ely, A., L.A. Berner, C.E. Wierenga, and W.H. Kaye. 2016. Neurobiology of eating disorders: Clinical implications. *Psychiatric Times* 33, no. 4. www.psychiatrictimes.com/view/neurobiology-eating-disorders-clinical-implications

Erskine, H.E., H.A. Whiteford, and K.M. Pike. 2016. The global burden of eating disorders. *Current Opinion in Psychiatry* 29, no. 6: 346–353.

Fairburn, C.G., H.A. Doll, S.L. Welch, P.J. Hay, B.A. Davies, and M.E. O'Connor. 1998. Risk factors for binge eating disorder: A community-based, case-control study. *Archives of General Psychiatry* 55, no. 5: 425–32. https://doi.org/10.1001/archpsyc.55.5.425

Fairburn, C.G., Z. Cooper, and R. Shafran. 2003. Cognitive behaviour therapy for eating disorders: A "trans-diagnostic" theory and treatment. *Behaviour Research and Therapy* 41: 509–528.

Fairburn, C.G., S. Bailey-Straebler, S. Basden, et al. 2015. A transdiagnostic comparison of enhanced cognitive behaviour therapy (CBT-E) and interpersonal psychotherapy in the treatment of eating disorders. *Behaviour Research and Therapy* 70: 64–71.

Fogarty, S., C.A. Smith, and P. Hay. 2016. The role of complementary and alternative medicine in the treatment of eating disorders: A systematic review. *Eating Behaviors* 21: 179–188.

Fogelkvist, M., S.A. Gustafsson, L. Kjellin, and T. Parling. 2020. Acceptance and commitment therapy to reduce eating disorder symptoms and body image problems in patients with residual eating disorder symptoms: A randomized controlled trial. *Body Image* 32: 155–166.

Frank, G.K.W., M.E. Shott, and M.C. DeGuzman. 2019. The Neurobiology of eating disorders. *Child and Adolescent Psychiatric Clinics of North America* 28, no. 4: 629–640. https://doi.org/10.1016/j.chc.2019.05.007.

Frostad, S., Y.S. Danielsen, G.A. Rekkedal, et al. 2018. Implementation of enhanced cognitive behaviour therapy (CBT-E) for adults with anorexia nervosa in an outpatient eating-disorder unit at a public hospital. *Journal of Eating Disorders* 6: 12. https://doi.org/10.1186/s40337-018-0198-y

Galmiche, M., P. Déchelotte, G. Lambert, and M.P. Tavolacci. 2019. Prevalence of eating disorders over the 2000–2018 period: A systematic literature review. *American Journal of Clinical Nutrition* 109, no. 5: 1402–1413. https://doi.org/10.1093/ajcn/nqy342

Ganci, M., L. Atkins, and M.E. Roberts. 2021. Exploring alternatives for adolescent anorexia nervosa: Adolescent and parent treatment (APT) as a novel intervention prospect. *Journal of Eating Disorders* 9: 67. https://doi.org/10.1186/s40337-021-00423-7

Golden, N.H., and W. Meyer. 2004. Nutritional rehabilitation of anorexia nervosa. Goals and dangers. *International Journal of Adolescent Medicine and Health* 16, no. 2: 131–144. https://doi.org/10.1515/ijamh.2004.16.2.131

Gorla, K., and M. Mathews. 2005. Pharmacological treatment of eating disorders. *Psychiatry* 2, no. 6: 43–48.

Ham, J.C., I. Daniela, and S. Michelle. 2021. Health outcomes, personality traits and eating disorders. *Economic Policy* 36, no. 105: 51–76. https://doi.org/10.1093/epolic/eiaa029

Herzog, D.B., and K.T. Eddy. 2009. Eating disorders: What are the risks?. *Journal of the American Academy of Child and Adolescent Psychiatry* 48, no. 8: 782–783. https://doi.org/10.1097/CHI.0b013e3181aa03d7

Hilbert, A., K.M. Pike, A.B. Goldschmidt, et al. 2014. Risk factors across the eating disorders. *Psychiatry Research* 220, no. 1–2: 500–506. https://doi.org/10.1016/j.psychres.2014.05.054

Holtkamp, K., B. Herpertz-Dahlmann, T. Vloet, and U. Hagenah. 2005. Group psychoeducation for parents of adolescents with eating disorders: The Aachen Program. *Eating Disorders* 13, no. 4: 381–390.

Hudson, J.I., E. Hiripi, H.G. Pope, and R.C. Kessler. 2007. The prevalence and correlates of eating disorders in the National Comorbidity Survey Replication. *Biological Psychiatry* 61, no. 3: 348–358. https://doi.org/10.1016/j.biopsych.2006.03.040

Jansingh, A., U.N. Danner, H.W. Hoek, and A.A. van Elburg. 2020. Developments in the psychological treatment of anorexia nervosa and their implications for daily practice. *Current Opinion in Psychiatry* 33, no. 6: 534–541.

Juarascio, A., J. Shaw, E. Forman, et al. 2013. Acceptance and commitment therapy as a novel treatment for eating disorders: An initial test of efficacy and mediation. *Behavior Modification* 37, no. 4: 459–489.

Juarascio, A.S., E.M. Forman, and J.D. Herbert. 2010. Acceptance and commitment therapy versus cognitive therapy for the treatment of comorbid eating pathology. *Behavior Modification* 34, no. 2: 175–190.

Kroger, C., U. Schweiger, V. Sipos, et al. 2010. Dialectical behaviour therapy and an added cognitive behavioural treatment module for eating disorders in women with borderline personality disorder and anorexia nervosa or bulimia nervosa who failed to respond to previous treatments. An open trial with a 15-month follow-up. *Journal of Behavior Therapy and Experimental Psychiatry* 41, no. 4: 381–388. https://doi.org/10.1016/j.jbtep.2010.04.001

Lammers, M.W., M.S. Vroling, R.D. Crosby, and T. van Strien. 2020. Dialectical behavior therapy adapted for binge eating compared to cognitive behavior therapy in obese adults with binge eating disorder: A controlled study. *Journal of Eating Disorders* 8: 27. https://doi.org/10.1186/s40337-020-00299-z

Levallius, J., D. Clinton, M. Backstrom, and C. Norring. 2015. Who do you think you are?-Personality in eating disordered patients. *Journal of Eating Disorders* 3: 3. https://doi.org/10.1186/s40337-015-0042-6

Loeb, K.L., and D. le Grange. 2009. Family-based treatment for adolescent eating disorders: Current status, new applications and future directions. *International Journal of Child and Adolescent Health* 2, no. 2: 243–254.

Lutter, M. 2017. Emerging treatments in eating disorders. *Neurotherapeutics* 14: 614–622. https://doi.org/10.1007/s13311-017-0535-x

Madden, S., S. Fogarty, and C. Smith. 2014. Alternative and complementary therapies in the treatment of eating disorders, addictions, and substance use disorders. In *Eating Disorders, Addictions and Substance Use Disorders*, edited by T. Brewerton and D.A. Baker. Springer, Berlin, Heidelberg. https://doi.org/10.1007/978-3-642-45378-6_29

Martini, M.G., M. Barona-Martinez, and N. Micali. 2020. Eating disorders mothers and their children: A systematic review of the literature. *Archives of Women's Mental Health* 23: 449–467.

Marzilli, E., L. Cerniglia, and S. Cimino. 2018. A narrative review of binge eating disorder in adolescence: Prevalence, impact, and psychological treatment strategies. *Adolescent Health, Medicine and Therapeutics* 9: 17–30.

Marzola, E., J.A. Nasser, S.A. Hashim, P.B. Shih, and W.H. Kaye. 2013. Nutritional rehabilitation in anorexia nervosa: Review of the literature and implications for treatment. *BMC Psychiatry* 13: 290. https://doi.org/10.1186/1471-244X-13-290

Mas, M.B., M.L.A. Navarro, A.M.L. Jimenez, I.T. Perez, C.D.R. Sanchez, and M.A.P.S. Gregorio. 2011. Personality traits and eating disorders: Mediating effects of self-esteem & perfectionism. *International Journal of Clinical and Health Psychology* 11, no. 2: 205–227.

Masuda, A., M.L. Hill, H. Melcher, J. Morgan, and M.P. Twohig. 2014. Acceptance and commitment therapy for women diagnosed with binge eating disorder: A case-series study. *Psychology Faculty Publications* 89. https://scholarworks.gsu.edu/psych_facpub/89

Mayhew, A.J., M. Pigeyre, J. Couturier, and D. Meyre. 2018. An evolutionary genetic perspective of eating disorders. *Neuroendocrinology* 106, no. 3: 292–306.

McKnight Investigators. 2003. Risk factors for the onset of eating disorders in adolescent girls: Results of the McKnight longitudinal risk factor study. *American Journal of Psychiatry* 160, no. 2: 248–54.

Mehler, P.S., A.B. Winkelman, D.M. Andersen, and J.L. Gaudiani. 2010. Nutritional rehabilitation: Practical guidelines for refeeding the anorectic patient. *Journal of Nutrition and Metabolism* 2010: 625782. https://doi.org/10.1155/2010/625782

Michael, J.E., C.M. Bulik, S.J. Hart, L. Doyle, and J. Austin. 2020. Perceptions of genetic risk, testing, and counseling among individuals with eating disorders. *International Journal of Eating Disorders* 53, no. 9: 1496–1505.

Milano, W., M.D. Rosa, L. Milano, B. Sanseverino, and A. Capasso. 2013. The pharmacological options in the treatment of eating disorders. *ISRN Pharmacology* 2013: 352865. https://doi.org/10.1155/2013/352865

Miniati, M., A. Callari, A. Maglio, and S. Calugi. 2018. Interpersonal psychotherapy for eating disorders: Current perspectives. *Psychology Research and Behavior Management* 11: 353–369. https://doi.org/10.2147/PRBM.S120584

Mudambi, S.R., and M.V. Rajagopal. 2009. *Fundamentals of Foods, Nutrition and Diet Therapy*. New Age International Publishers, New Delhi, 350–352.

Mulders-Jones, B., D. Mitchison, F. Girosi, and P. Hay. 2017. Socioeconomic correlates of eating disorder symptoms in an Australian population based sample. *Plos One* 12, no. 1: e0170603. https://doi.org/10.1371/journal.pone.0170603

Murphy, R., S. Straebler, S. Basden, Z. Cooper, and C.G. Fairburn. 2012. Interpersonal psychotherapy for eating disorders. *Clinical Psychology and Psychotherapy* 19, no. 2: 150–158.

Murphy, R., S. Straebler, Z. Cooper, and C.G. Fairburn. 2010. Cognitive behavioral therapy for eating disorders. *Psychiatric Clinics of North America* 33, no. 3: 611–627.

Neugebauer, Q., S. Mack, A. Roubin, and A. Curiel. 2011. Primary prevention of eating disorders in children and a proposed parent education program. *Graduate Student Journal of Psychology* 13: 52–59.

Nicholls, D.E., and I. Yi. 2012. Early intervention in eating disorders: A parent group approach. *Early Intervention in Psychiatry* 6, no. 4: 357–367. https://doi.org/10.1111/j.1751-7893.2012.00373.x

Nimrouzi, M., and M.M. Zarshenas. 2018. Anorexia: Highlights in traditional Persian medicine and conventional medicine. *Avicenna Journal of Phytomedicine* 8, no. 1: 1–13.

Pathak, P., V. Kukreti, A.T. Bisht, and P.D. Bhatt. 2020. A profile of depression, mental health, BMI and food avoidance by school going adolescent girls. *Our Heritage* 68, no. 14: 487–493.

Pathak, P., V. Kukreti, and A.T. Bisht. 2019. Parental perceptions, depression and BMI: A road to understand the occurrence of vulnerability to anorexia in adolescent girls. *Online Journal of Health and Allied Sciences* 18, no. 3: 7. www.ojhas.org/issue71/2019-3-7.html

Pike, K.M., and P.E. Dunne. 2015. The rise of eating disorders in Asia: A review. *Journal of Eating Disorders* 3: 33. https://doi.org/10.1186/s40337-015-0070-2

Qian, J., Y. Wu, F. Liu, et al. 2022. An update on the prevalence of eating disorders in the general population: A systematic review and meta-analysis. *Eating and Weight Disorders* 27: 415–428. https://doi.org/10.1007/s40519-021-01162-z

Rahmani, M., A. Omidi, Z. Asemi, and H. Akbari. 2018. The effect of dialectical behaviour therapy on binge eating, difficulties in emotion regulation and BMI in overweight patients with binge-eating disorder: A randomized controlled trial. *Mental Health & Prevention* 9: 13–18.

Rieger, E., D.J. Van Buren, M. Bishop, M. Tanofsky-Kraff, R.Welch, and D.E. Wilfley. 2010. An eating disorder-specific model of interpersonal psychotherapy (IPT-ED): Causal pathways and treatment implications. *Clinical Psychology Review* 30, no. 4: 400–410.

Rienecke, R.D. 2017. Family-based treatment of eating disorders in adolescents: Current insights. *Adolescent Health, Medicine and Therapeutics* 8: 69–79. https://doi.org/10.2147/AHMT.S115775

Rikani, A.A., Z. Choudhry, A.M. Choudhry, et al. 2013. A critique of the literature on etiology of eating disorders. *Annals of Neurosciences* 20, no. 4: 157–61. https://doi.org/10.5214/ans.0972.7531.200409

Rizzuto, L., P. Hay, M. Noetel, and S. Touyz. 2021. Yoga as adjunctive therapy in the treatment of people with anorexia nervosa: A Delphi study. *Journal of Eating Disorders* 9, no. 1: 111. https://doi.org/10.1186/s40337-021-00467-9

Rohde, P., E. Stice, H. Shaw, J.M. Gau, and O.C. Ohls. 2017. Age effects in eating disorder baseline risk factors and prevention intervention effects. *International Journal of Eating Disorders* 50, no. 11: 1273–1280. https://doi.org/10.1002/eat.22775

Rome, E.S. and S. Ammerman. 2003. Medical complications of eating disorders: An update. *Journal of Adolescent Health* 33, no. 6: 418–426.

Safer, D.L., C.F. Telch, and W.S. Agras. 2001. Dialectical behavior therapy for bulimia nervosa. *American Journal of Psychiatry* 158, no. 4: 632–634.

Santomauro, D.F., S. Melen, D. Mitchison, T. Vos, H. Whiteford, and A.J. Ferrari. 2021. The hidden burden of eating disorders: An extension of estimates from the Global Burden of Disease Study 2019. *Lancet Psychiatry* 8, no. 4: 320–328. https://doi.org/10.1016/S2215-0366(21)00040-7

Shindhe, S.S., N. Kusuma, Nagarajaiah, and B.M. Suresh. 2014. Psychoeducation for mental illness: A systematic review. *Indian Journal of Psychiatry Nursing* 8: 46–52.

Shumlich, E.J. 2017. Dialectical behaviour therapy and acceptance and commitment therapy for eating disorders: Mood intolerance as a common treatment target. *Canadian Journal of Counselling and Psychotherapy* 51, no. 3: 217–229.

Smink, F.R., D. van Hoeken, and H.W. Hoek. 2012. Epidemiology of eating disorders: Incidence, prevalence and mortality rates. *Current Psychiatry Reports* 14, no. 4: 406–414.

Solmi, M., J. Radua, B. Stubbs, et al. 2021. Risk factors for eating disorders: An umbrella review of published meta-analyses. *Brazilian Journal of Psychiatry* 43: 314–323.

Spettigue, W., D. Maras, N. Obeid, et al. 2015. A psycho-education intervention for parents of adolescents with eating disorders: A randomized controlled trial. *Eating Disorders* 23, no. 1: 60–75.

Steinglass, J.E., L.A. Berner, and E. Attia. 2019. Cognitive neuroscience of eating disorders. *Psychiatric Clinics of North America* 42, no. 1: 75–91.

Storch, M., F. Keller, J. Weber, A. Spindler, and G. Milos. 2011. Psychoeducation in affect regulation for patients with eating disorders: A randomized controlled feasibility study. *American Journal of Psychotherapy* 65, no. 1: 81–93.

Striegel-Moore, R.H., and C.M. Bulik. 2007. Risk factors for eating disorders. *American Psychologist* 62, no. 3: 181–198.

Striegel-Moore, R.H., F. Rosselli, N. Perrin, et al. 2009. Gender difference in the prevalence of eating disorder symptoms. *International Journal of Eating Disorders* 42, no. 5: 471–474.

Tanofsky-Kraff, M., D.E. Wilfley, J.F. Young, et al. 2010. A pilot study of interpersonal psychotherapy for preventing excess weight gain in adolescent girls at-risk for obesity. *International Journal of Eating Disorders* 43, no. 8: 701–706. https://doi.org/10.1002/eat.20773

Trace, S.E., J.H. Baker, E. Peñas-Lledó, and C.M. Bulik. 2013. The genetics of eating disorders. *Annual Review of Clinical Psychology* 9: 589–620. PMID: 23537489. https://doi.org/10.1146/annurev-clinpsy-050212-185546

Uehara, T., Y. Kawashima, M. Goto, S.I. Tasaki, and T. Someya. 2001. Psychoeducation for the families of patients with eating disorders and changes in expressed emotion: A preliminary study. *Comprehensive Psychiatry* 42, no. 2: 132–138.

Upadhyah, A.A., R. Misra, D.N. Parchwani, and P.B. Maheria. 2014. Prevalence and risk factors for eating disorders in Indian adolescent females. *National Journal of Physiology, Pharmacy and Pharmacology* 4: 153–157.

van Eeden, A.E., D. van Hoeken, and H.W. Hoek. 2021. Incidence, prevalence and mortality of anorexia nervosa and bulimia nervosa. *Current Opinion in Psychiatry* 34, no. 6: 515–524.

van Hoeken, D., and H.W. Hoek. 2020. Review of the burden of eating disorders: Mortality, disability, costs, quality of life, and family burden. *Current Opinion in Psychiatry* 33, no. 6: 521–527. https://doi.org/10.1097/YCO.0000000000000641.

Ward, Z.J., P. Rodriguez, D.R. Wright, S.B. Austin, and L.W. Long. 2019. Estimation of eating disorders prevalence by age and associations with mortality in a simulated nationally representative US cohort. *JAMA Network Open* 2, no. 10: e1912925. https://doi.org/10.1001/jamanetworkopen.2019.12925

Wild, B., H.C. Friederich, G. Gross, et al. 2009. The ANTOP study: Focal psychodynamic psychotherapy, cognitive-behavioural therapy, and treatment-as-usual in outpatients with anorexia nervosa--A randomized controlled trial. *Trials* 10: 23. https://doi.org/10.1186/1745-6215-10-23

Wilson, G.T., D.E. Wilfley, W.S. Agras, and S.W. Bryson. 2010. Psychological treatments of binge eating disorder. *Archives of General Psychiatry* 67, no. 1: 94–101.

Wu, J., J. Liu, S. Li, H. Ma, and Y. Wang. 2020. Trends in the prevalence and disability-adjusted life years of eating disorders from 1990 to 2017: Results from the global burden of disease study 2017. *Epidemiology and Psychiatric Sciences* 29: E191. https://doi.org/10.1017/S2045796020001055

www.ucl.ac.uk. Focal psychodynamic therapy for anorexia nervosa. www.ucl.ac.uk/clinical-psychology/competency-maps/eating-disorders

Yilmaz, Z., J.A. Hardaway, and C.M. Bulik. 2015. Genetics and epigenetics of eating disorders. *Advances in Genomics and Genetics* 5: 131–150.

Zipfel, S., B. Wild, G. Groß, et al. 2014. Focal psychodynamic therapy, cognitive behaviour therapy, and optimised treatment as usual in outpatients with anorexia nervosa (ANTOP study): Randomised controlled trial. *Lancet* 383, no. 9912: 127–137.

9 Depression

9.1 INTRODUCTION

Mental health is an integral part of holistic health. It is one of the essential components of health along with physical and social health. Mental illness is characterised by a change in thinking, emotions, and behaviour. It is associated with functional impairment in social, work, or family activities (Parekh 2018). According to the World Health Organization (WHO), mental health is defined as a state of wellbeing in which a person realises their own abilities, copes with the normal stressors in life, works productively, and makes a contribution to their community (PAHO nd). Mental illness impacts one's abilities to handle stress, make healthy choices, and relate to others as it affects the thinking, emotions, and behaviour of an individual (CDC 2021). Mental illnesses are highly prevalent in the current era and contribute to morbidities, disabilities, and premature mortalities. They act as a risk factor for the development of other diseases and also lead to intentional or unintentional injuries (PAHO nd).

Various organisations like the WHO and the American Psychological Association (APA), have classified mental disorders and given a broad spectrum of mental illnesses. However, this chapter focuses on depression. Depression is a mental health issue confronted by each individual at some point in life, and coping with it often becomes difficult, causing a major health challenge. The rapid transition in lifestyle and social functioning has aggravated depressed mental states. Also, depression is linked with many other psycho-physiological health conditions and therefore it becomes important to converse about depression in detail.

Depression is an affective disorder characterised by pathological changes in mood. Depression influences the way we think, feel, and behave. It is a mood disorder that causes persistent feelings of sadness, helplessness, losing interest in people around us, and in pleasurable activities.

9.2 SYMPTOMS OF DEPRESSION

The common signs and symptoms of depression (WHO 2021a) are as follows:

- Feeling sad, irritable, empty
- Loss of pleasure
- Loss of interest
- Change in weight
- Change in sleep
- Change in appetite
- Poor concentration
- Fatigue
- Weakness
- Bodily pain
- Feeling of guilt
- Feeling of low self-worth
- Feeling of hopelessness
- Suicidal thoughts

DOI: 10.1201/9781003354024-9

These symptoms are categorised as core symptoms, somatic symptoms, and other symptoms to ease the diagnosis and to identify the severity of symptoms shown in Figure 9.1 (National Collaborating Centre for Mental Health 2010).

9.3   DIAGNOSIS OF DEPRESSION

According to Diagnostic and Statistical Manual of Mental Disorders (DSM) -5 (APA 2013), diagnosis of major depression requires the presence of five or more symptoms, of which either one or both symptoms should be from the core symptom category, i.e., depressed mood and/or anhedonia, and the rest from the secondary symptoms. The presence of three or more non-somatic symptoms like feelings of guilt, impaired concentration, and suicidal ideation, along with core symptoms, becomes a quick guide for practitioners to diagnose severe depression. However, the importance of somatic symptoms in determining the severity of depression cannot be overlooked, particularly if they are not an outcome of some other physical illness (National Collaborating Centre for Mental Health 2010). Among the core symptoms, anhedonia is a major symptom for severe depression and can be used to discriminate between severe and moderate depression, whereas depressed mood is an important symptom to discriminate between moderate depression and non-depressed subjects. Among the secondary symptoms, the presence of somatic symptoms discriminates moderately depressed individuals from non-depressed individuals, and non-somatic symptoms discriminate severe depression from moderate depression (Tolentino and Schmidt 2018)

9.4   CLASSIFICATION OF DEPRESSION

Depression can be classified depending on the types of symptoms, their severity, and their frequency of occurrence. Depression can be classified as subthreshold depressive symptoms, mild, moderate, and severe depression depending on the severity of symptoms (National Collaborating Centre for Mental Health 2010).

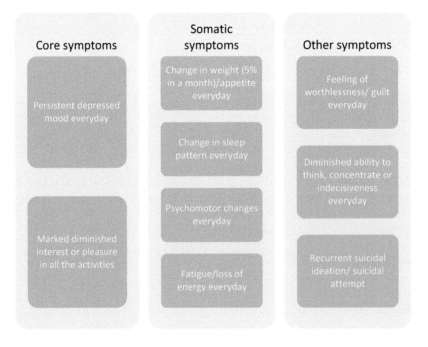

FIGURE 9.1   Categorisation of depressive symptoms.

- Subthreshold depressive symptoms: fewer than five symptoms of depression
- Mild depression: when more than five symptoms are required to make the diagnosis and the symptoms result in minor functional impairment
- Moderate depression: when symptoms and functional impairment are between mild and severe
- Severe depression: when symptoms interfere with functioning

Depression can be classified as acute or chronic on the basis of the persistence of symptoms for a length of time.

- Acute: when the symptoms persist for more than two weeks
- Chronic: when the symptoms persist for longer than two years

9.5 PREVALENCE OF DEPRESSION

Depression is a major public health problem and leading cause of disability. Major depressive disorders were responsible for 2.5% of global disability adjusted life years (DALYs) and dysthymia for 0.5% (Ferrari et al. 2013). According to the findings from the Global Burden of Diseases, injuries and risk factor study 2010, amongst the diseases attributable to mental and substance use disorders, depression alone accounted for 40.4% of Disability-adjusted life years (DALYs) (Whiteford et al. 2013).

Liu et al. (2020) analysed Global Burden of Disease data and concluded an increase of almost 50% in incidences of depression in three decades. They inferred that worldwide the number of cases of depression increased from 172 million in 1990 to 258 million in 2007, of which 241 million cases were for major depressive disorder and 16 million for dysthymia. Presently, approximately 280 million people have depression affecting 3.8% of the population. It is estimated that 5% of adults and 5.7% of adults older than 60 years are suffering from depression (WHO 2021a).

In recent times, the COVID-19 pandemic has further aggravated the situation of depression globally. The COVID-19 Mental Disorders Collaborators (2021) analysed the severity of depression before and after the pandemic and found an increase in the prevalence of depression. In 2020, before COVID, the estimated global prevalence of depression was estimated to be 2470.5 cases per 100,000 population (193 million), which has increased to 3159.2 cases per 100,000 population (246 million), representing an upsurge of 53.2 million cases due to COVID-19.

The prevalence of major depressive disorder across the world showed the highest prevalence in European countries and lowest in the Asia region (Gutierrez-Rojas et al. 2020). Females are more susceptible to depression. However, the gender gap is greatest at adolescence, which gradually narrows down and becomes stable in adulthood (Salk et al. 2017). Besides female gender, unemployment status, education level, age, place of residence, and household income are associated with a higher prevalence of major depressive disorder (Gutierrez-Rojas et al. 2020; Arvind et al. 2019).

9.6 AETIOLOGY OF DEPRESSION

Depression has a multi-factorial aetiology, with genetics and environmental factors playing pivotal roles in the development of depression. Besides, many factors like developmental stage, hormonal transition, and gender act as moderators in pathogenesis of depression. Himmerich et al. (2019) have pointed out the following risk factors in the development of depression:

i. Adolescent age
ii. Psychosocial risk factors
iii. Oxidative stress
iv. Nutrition and the gut microbiome

v. Psychotic disorders
vi. Brain diseases
vii. Physical diseases
viii. Family history
ix. Hormonal changes
x. Disturbed sleep pattern
xi. Genetic risk factors
xii. Functional and structural changes in the brain
xiii. Neurochemical risk factors

The risk factors contributing to depression can be broadly categorised as biological, environmental, and personal vulnerabilities shown in Figure 9.2.

9.6.1 Biological Factors

The biological factors include genetic, neurological, hormonal, and neuroendocrinological mechanisms in the development of depression (National Research Council and Institute of Medicine 2009). Past studies have shown that depression is heritable. Parental 'depression genetic makeup' may increase the susceptibility of developing depression in offspring. Family, twin, and adoption studies have supported that depression runs in families, and persons with depression are 2–3 times more likely to have a first-degree relative with depression (Sullivan et al. 2000) and heritability of depression can be as high as 42% for women and 29% for men (Kendler et al. 2006). The heritability of depression was estimated at 28% after adjusting for shared family environment. Sex, age of onset, and course of illness were correlated with heritability, and a strong positive genetic correlation between major depressive disorders in males and females, earlier and later onset of depression, and single and recurrent episodes of depression were found. This indicates the role of genes in developing depression (Fernandez-Pujals et al. 2015). Genome-wide association studies (GWAS) have identified genetic variants associated with depression (Mullins and Lewis 2017). Genome-wide linkage scan analysis of families with multiple cases of recurrent early-onset major depressive disorders suggested chromosomes 15q, 17p, and 8p might contain genes that contribute to the high

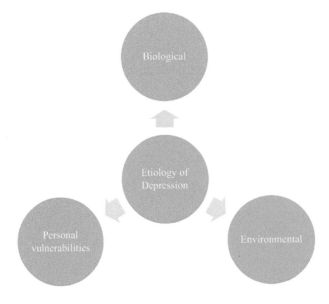

FIGURE 9.2 Aetiological factors contributing to depression.

susceptibility, indicating multiple loci may contribute to the risk of developing depression (Holmans et al. 2007).

Other biological mechanisms associated with depression are insufficiency of monoamine neuro-mediators (Shadrina et al. 2018), elevated cortisol level (Qin et al. 2016), increase in inflammation, dysregulation of the hypothalamic-pituitary-adrenal (HPA) axis, lower vitamin D status (Verduijn et al. 2015) and suppressed brain-derived neurotrophic factor (BDNF) synthesis (Palazidou 2012). It has been established that stress, along with genetic diathesis, influences the biological mechanism, which consequently acts as a determinant of depression onset. Depression is caused by a deficiency of monoamines, noradrenaline and serotonin, in the brain (Ruhe et al. 2007). The hippocampus is rich in BDNF, which plays a pivotal role in neuronal growth, survival, and maturation. It is also important for synaptic plasticity, but stress suppresses BDNF synthesis (Palazidou 2012). Morphological changes in the hippocampus may be a predisposing factor in major depressive disorder. The hormone cortisol is released in elevated levels because of the failure to withstand the stressors (Maletic et al. 2007). These interlinked biological processes compromise neurogenesis and neuroplasticity and cause aberrations in the neuronal network and plasticity, which lead to depression.

Further understanding the phenomena, the brain responds to stress by activation of the HPA axis wherein the hypothalamus secretes corticotropin-releasing factor (CRF). CRF stimulates the synthesis and release of cortisol. Cortisol, especially in persistent elevated levels, damages hippocampal neurones, reduces the birth of new neurones, thereby impacting the general metabolism and behaviour (Nestler et al. 2002).

The field of genetics is quite complex and research is still ongoing to establish the precise relationship. However, the role of environment cannot be ignored. A multitude of studies has inferred that depression is a product of gene and environment interaction. Each susceptible gene itself has a very low contribution in inducing depression. It is the interaction of multiple and partially overlapping genes in conjunction with environmental factors that determine the risk of developing depression (Lopizzo et al. 2015; Fava and Kendler 2000).

9.6.2 ENVIRONMENTAL FACTORS

Environmental factors, along with biological factors and personal vulnerabilities of an individual, increase the sensitivity towards developing depression. Adverse environmental factors enhance the chances of developing depression more in genetically susceptible individuals. Environmental factors, comprising of physical environmental factors and social environmental factors, are the external factors that interfere with the normal neurobiological functioning, increasing the risk of developing depression. Environmental factors contribute to chronic stress which subsequently alters the neurobiological mechanism leading to depression.

The physical environmental factors may be categorised into three broad groups depending upon their proximity with an individual's brain and central nervous system (CNS). Immediate physical environmental factors, such as air pollutants, pollen, noise, BPA/phthalates, pesticides, and heavy metals; proximate contextual environmental factors, such as family, friends, neighbourhood, and workplace; and the distant contextual physical environment, such as urbanicity, political system, economics, war/conflicts, climate change, and disasters. These physical components, alone or in combination with other factors, affect the normal functioning and structure of the brain and CNS (van den Bosch and Meyer-Lindenberg 2019). Exposure to harmful physical environments causes cell damage and impairs the functioning of the CNS, contributing to the development of depression.

Airborne pollutants emitted from vehicles, industry, and other sources enter the body through inhalation or the skin. It has been reported that fine particulate matter (<2.5 micron) enters the circulation and reaches the brain directly, affecting the functioning of the CNS and neurotransmitters that are associated with the generation of depression. Besides, they may induce oxidative stress, influence neural plasticity, and affect neuronal morphology (van den Bosch and Meyer-Lindenberg 2019).

Exposure to traffic-related air pollution is directly correlated with depressive disorders (Pelgrims et al. 2021).

Exposure to aeroallergens is associated with an increase in inflammatory cytokines and CNS cholinergic system sensitivity, which consequently induce depression (van den Bosch and Meyer-Lindenberg 2019). Worsening depression scores wereobserved when exposed to pollens, especially in allergen-specific Immunoglobulin E (IgE)-positive subjects (Manalai et al. 2012). It has been seen that an increase in air pollen count increased the risk of suicide, particularly in men and those with mood disorders (Qin et al. 2013).

Noise is another physical environmental factor that has an impact on the HPA axis (van den Bosch and Meyer-Lindenberg 2019). It was observed that the risk of depression increased two-fold with the increase in noise annoyance (Beutel et al. 2016).

In recent years, exposure to human-made electromagnetic fields has been increasing alarmingly, putting all the age groups at risk. The radiation generated from human-made electromagnetic fields may damage the brain and neuronal structures, along with affecting the neurotransmitter system (van den Bosch and Meyer-Lindenberg 2019). Exposure to extremely low frequency electromagnetic fields increased the severity of depression (Bagheri Hosseinabadi et al. 2019). Smart phone addiction was associated with depression, especially in children and young adults (Alhassan et al. 2018; Sohn et al. 2019). It has been observed that an increase in screen time (usage of social media, television, and computers) was associated with an increase in depressive symptoms (Boers et al. 2019)

Heavy metals like cadmium, lead, and mercury are neurotoxic, and endocrine-disrupting substances like bisphenol, phthalates, and pesticides disrupt the normal functioning of the brain, causing neural degeneration, hyperactivity of the HPA axis, interference with synaptogenesis, and increased vulnerability in children and during prenatal developmental stages (van den Bosch and Meyer-Lindenberg 2019).

Drugs like immunomodulatory, hormonal, antihypertensive, psychotropic may cause depression (Nabeshima and Kim 2013). Certain drugs such as β-blockers, digoxin, and steroids may induce depression, more so in the subjects with a history of depression or a family history of depression (Patten and Love 1993). Antiobesity, dermatologics, antiviral, corticosteroids, and smoking cessation agents have been associated with the induction of depression (Rogers and Pies 2008). However, not very conclusive results have been drawn by the researchers.

Besides physical environmental factors, certain psychosocial factors also play a vital role in causing depression. Social support received inside and outside the family is associated with depression (Yan et al. 2022). It has been seen that social support received from family and friends was protective against depression (Ioannou et al. 2019). Social support enhances resilience to stress possibly through its impact on the HPA axis, noradrenergic system, and central oxytocin pathways (Ozbay et al. 2007). Experiencing household dysfunction during childhood is a risk factor for developing depression during adolescence, indicating the role of social support in mitigating depression (Tsehay et al. 2020). Childhood trauma increases vulnerability towards the development of depression during adulthood. Trauma of any kind, such as neglect, emotional abuse, sexual abuse, or physical abuse experienced by a child, increases the chances of developing chronic depression in adulthood (Negele et al. 2015). Childhood trauma, except physical, leads to interpersonal distress in adulthood (Huh et al. 2014). Childhood trauma affects brain networking. Childhood traumatic experiences correlate with disruption of network connectivity within various regions of the brain and sensory system, indicating childhood maltreatment to have an influence on brain morphology and functioning (Yu et al. 2019). Childhood experiences of abuse have a long-lasting effect, to the extent that such individuals respond poorly to treatment during adulthood and have a higher suicidal tendency. The extent could be estimated from the results that showed that maternal childhood traumatic experience increases the risk of developing depression in offspring (Wagner 2016). Family support is very important. Family environment factors such as poor parent relationship, poor family economic status, low level of parental literacy, conflict, control, and lack of family cohesion were seen to be associated with depression (Yu et al. 2015). Family biosocial variables like unhealthy family functioning, low family expressiveness, and high family conflict are related to depression. It has been observed that unhealthy family functioning increases the likelihood of depression by three times compared to subjects with healthy family functioning (Pascal Iloh et al. 2018).

Neighbourhood also has a direct impact on the mental wellbeing of an individual. An unfavourable neighbourhood, firstly in terms of the level of daily stress influenced by characteristics of the neighbourhood; secondly, triggering vulnerability to any negative event faced by an individual; and thirdly, disrupted social ties with neighbours have an influence on stress leading to depression. The poor structural and functional neighbourhood characteristics are directly associated with depression. It has been reported that the stress imposed by adverse neighbourhood increases depression above and beyond the effect of personal stressors, indicating the impact of the neighbourhood in triggering depressive disorders (Cutrona et al. 2006).

Biological, chemical, physical, social, and psychological stressors creating a negative occupational environment lead to job-related depression (Woo and Postolache 2008). It has been reported that an adverse psychosocial work environment, as measured in terms of demand control (job strain), effort-reward imbalance, and organisational injustice, was related to an increase in depressive symptoms (Siegrist and Wege 2020). Thus, family, friends, neighbourhood, and workplace which constitute our proximate contextual environment, have an impact on triggering depression.

Distant contextual environments such as war and conflict, disasters, and climate change have a catastrophic effect on mental health. During war, conflict, and disasters, women, elderly, handicapped, and children are affected the most (Murthy and Laxminarayana 2006; Makwana 2019). The physical, psychological, social, and ecological trauma experienced during wars leads to depression (Musisi and Kinyanda 2020). Similarly, it has been inferred that disasters of any kind, i.e., man-made or natural, lead to depression in survivors. Disasters bring the victims into a state of shock, increase stress and depression, along with other psychological problems (Makwana 2019; Morganstein and Ursano 2020). The recent example is COVID-19 where various studies showed the prevalence of depression to vary from 33.7% (Salari et al. 2020) to 34.3% and was comparatively higher in China (36.32%) compared to other countries (28.3%) (Necho et al. 2021). This indicates a rise in the prevalence of depression. during the pandemic compared to normal times (Mahmud et al. 2022). Climate change also has an impact on the depressive state. Extreme events such as an increase in temperature, heatwaves, floods, droughts, tornadoes, hurricanes, and wildfires, as a result of global climate change, increase the rate of depression (Cianconi et al. 2020). Climate change affects climatic elements such as the quality of rainfall, low sunshine, seasonal changes, and cloudiness, leading to depression (Abbasi 2021). Events like floods, droughts, famines, tornadoes, and hurricanes that originate due to climatic changes cause economic strain, forceful migration, acculturation stress, and multiple traumas, leading to depression and other mental health disorders (Padhy et al. 2015).

A few more environmental factors have been found to be associated with depression. Poor indoor physical environment such as improper available basic home facilities, ventilation, and sunny environment, home satisfaction, neighbourhood physical environment, and work physical environment were seen to be associated with depression (Ragab et al. 2000). Urbanisation, though not always, is a reason for mental illness, especially depression. It has been observed that a rural to urban shift sometimes causes socioeconomic stress, cultural transformation shock, an increase in domestic violence particularly for women, exposure to a poor physical environment such as less greenery, more pollution, urban density (Srivastava 2009; Mc Eachan et al. 2016; Melis et al. 2015).

Environmental factors like *in utero* exposure to infection, lack of nutrients, maternal stress, prenatal complications, social disadvantage, urban upbringing, ethnic minority status, and cannabis use increase the vulnerability towards the development of depression (Lopizzo et al. 2015).

Patients suffering from chronic and degenerative diseases are likely to get depressed with the severity of illness (Nabeshima and Kim 2013; Dagnino et al. 2020). All of the above-stated environmental conditions are risk factors in predisposing depression. Environmental factors alter the neural oscillatory activity of the cerebellum during the resting state thereby increasing susceptibility towards the development of depression (Cordova-Palomera 2016).

9.6.3 Personal Vulnerabilities

Despite the same degree of genetic and environmental exposure, the susceptibility to developing depression varies from individual to individual. The personal traits of an individual mediate the response towards the same exposure. Personal vulnerabilities such as cognitive vulnerabilities, interpersonal vulnerabilities, and specific personality traits increase the risk of developing depression. Cognitive vulnerabilities are beliefs and attitudes of an individual that make him vulnerable to depression. Cognitive characteristics include poor self-esteem, low social competence, and negative attributes that increase the likelihood of getting depressed. Interpersonal vulnerabilities represent poor interaction with others that may ultimately affect the person. For example, marital discord, intimate partner violence, parenting difficulties, insecure attachment, and low social support. Personality or temperament traits refer to a person's reaction and behaviour to an event. For instance, neuroticism is a temperament trait (National Research Council (US) and Institute of Medicine (US) Committee 2009).

These three broad components of personal vulnerability are intertwined. There is an interrelationship between cognitive vulnerability, temperament traits, and interpersonal relationships. Temperament traits and interpersonal relationships influence the cognitive vulnerability to depression. It has been stated that an individual with a susceptible gene for a particular temperament may evoke negative interaction from a weak interpersonal relationship, leading to cognitive vulnerability towards depression. Thus, the temperament traits which are heritable play a role in the development of cognitive vulnerabilities. Three major temperament traits, negative emotionality, positive emotionality, and attentional control, are related to depression. High negative emotionality and low positive emotionality and attentional control contribute to depression either by generating stress or reducing social support or moderating other temperament traits. Temperament, poor interpersonal relationships, and stressful events lead to dysfunctional attitudes and a negative attributional style indicative of cognitive vulnerability to depression. High criticism, rejection, control, and low warmth and acceptance increase cognitive vulnerability to depression (Hankin et al. 2009). For instance, individuals with personality traits of dependency develop depression during a negative interpersonal event such as personal loss or social rejection or disrupted relations. Similarly, an individual with self-criticism personality trait develops depression during a negative achievement event (Abela and Hankin 2008).

Interpersonal relationships are the relationships between individuals and expressions of the reactions and emotions of each individual to others. Common interpersonal relationships include relations within the family, the social environment, and interactions between genders. Any kind of discord in familial relationships, social rejection, or inability to maintain social norms, especially in the case of females, may lead to depression (Beattie 2005). Helplessness in maintaining healthy social relationships, probably due to a lack of interpersonal skills, leads to hopelessness and loneliness, which ultimately trigger depression (Ivano 2005). Depression can be triggered by negative social experiences, more for those individuals who cannot cope or handle them due to their personal vulnerabilities. The likelihood of developing depression varies from person to person depending on how the individual process these experiences.

The temperament profile of an individual directly or indirectly (by influencing cognitive vulnerability) contributes to depression. A multitude of studies has been conducted to identify the temperament variables associated with depression by comparing the temperament profile of subjects suffering from depression with their healthy counterparts, and it was seen that depressed subjects had higher novelty seeking, harm avoidance, self-transcendence, and lower reward dependence, persistence, self-directedness, and cooperativeness traits than their non-depressed equivalents (Nery et al. 2009; Jylha et al. 2011; Zappitelli et al. 2013; Nogueira et al. 2017). Higher personal reserves, rejection sensitivity, and self-criticism are correlated with a higher level of depression (Kudo et al. 2017). These negative temperament traits affect the intensity of depressive symptoms, remittance episodes, and treatment processes.

Temperament and early childhood adverse experiences trigger early maladaptive schemas, which hamper the coping styles against depression (Lim et al. 2018). Low sociability and high negative emotionality increase depression symptoms (Elovainio et al. 2015).

Research shows the variables associated with depression in varying populations. Frustration (Oldehinkel et al. 2006), pessimism, negative attitude towards self, self-criticism, low optimism (Balsamo et al. 2013), hopelessness, worthlessness (Marchetti et al. 2016), poor resilience (Haddadi and Besharat 2010), sociotropic, autonomous personality dimensions (Frewen et al. 2008), other personality dimensions such as neuroticism, extraversion, dependency, self-criticism, melancholy, obsessionality, perfectionism, negative emotionality, high interpersonal sensitivity (Boyce et al. 1991; Enns and Cox 1997; Klein et al. 2011; Nabeshima and Kim 2013), increased negative affectivity (Echezarraga et al. 2021), type D personality which is a combination of negative affectivity and social inhibition (van Dooren et al. 2016), low self-directedness (Andriola et al. 2011) are associated with depression. A structure of temperament questionnaire was deployed on depressed and non-depressed subjects ranging in age from 17–84, and it was seen that depressed subjects scored higher on impulsivity and neuroticism and lower on motor-physical endurance, social endurance, social tempo, intellectual endurance, plasticity, and self-confidence (Trofimova and Sulis 2016).

Less integrated personality functioning were observed to be associated with depression symptomatology (Dagnino et al. 2020). Negative interpersonal relationships with family, siblings, and friends or in the the workplace increase loneliness and social exclusion (Majd Ara et al. 2017; Stoetzer et al. 2009). Poor social skills lead to depression (Segrin 2000).

9.7   PATHOGENESIS OF DEPRESSION

As we know, biological mechanisms (genetic diathesis) and environmental factors, along with personal vulnerabilities, lead to the development of depression. Therefore, it is important to understand the interplay of these factors in the pathogenesis of depression.

It is well researched that insufficiency or imbalance of neuromediators like serotonin, norepinephrine, or dopamine leads to depression and this is because of the multiple stressors encountered by an individual. Stressors lead to dysfunctioning in the serotonin system such as reduced concentration, impaired uptake, altered binding, or tryptophan depletion which is correlated with depression. The serotonin system imbalance also affects the functioning of the HPA axis and other brain regions (National Research Council and Institute of Medicine Committee, 2009). The hyperactivity of the HPA system due to stress initiates depression. A decrease in the normal neurogenesis process attributed to disturbance in BDNF in nervous tissue may lead to the development of depression as well (Shadrina et al. 2018).

Inflammatory cytokines generated from stress also play a role in the induction of depression. Stress-driven inflammatory cytokines affect the activity of the HPA axis through multiple mechanisms. They decrease neurogenesis, increase neurotoxicity, decrease serotonin level, increase CRH and lower BDNF (Madeeh Hashmi et al. 2013). Cytokines stimulate astrocytes to release glutamate, which leads to a decrease in BDNF, which is indispensable for neurogenesis. Cytokines affect serotonin metabolism and signalling. euroendocrine functioning is affected by cytokines by increasing HPA axis activity, CRH, adrenocorticotropic hormone (ACTH), and cortisol (Himmerich et al. 2019). These dysregulations affect the brain regions. Neuroimaging studies have found aberrant structure, functioning, and connectivity in brain regions related with emotional regulation in depressed subjects. Individuals with depression showed structural abnormalities in the amygdala and hippocampus and functional abnormalities in the subgenual anterior cingulate cortex, dorsolateral prefrontal cortex, amygdala, and ventral striatum (Singh and Gotlib 2014). A reduction in brain volume especially in the anterior cingulate, the putamen, and the caudate, has been seen. Also, abnormalities in regional cerebral blood flow and glucose metabolism in multiple prefrontal cortical and limbic structures implicated in emotional processing are seen in depressed subjects. Decreased activity of the dorsolateral prefrontal cortex is associated with psychomotor retardation and anhedonia. There is an intricate limbic-cortico-striato-pallido-thalamic neurocircuit that is responsible for maintaining emotional stability and regulating endocrine function. It has been observed that during a depressed state, the functioning of this complex neurocircuit becomes disrupted, and this leads to depression (Palazidou 2012).

Palazidou (2012) has explained the pathogenesis of depression in quite comprehensive and simple terms. Genetic diathesis and stress are responsible for a lowered level of serotonin and noradrenaline transmission and an increase in proinflammatory cytokines which further cause a decrease in BDNF and HPA dysregulation, leading to hippocampal dystrophy followed by prefrontal cortex hypoactivity and limbic hyperactivity. This mechanism accounts for the development of depression. The ventral anterior cingulate cortex is involved in assessing emotional and motivational information, and the dorsal anterior cingulate cortex is associated with cognitive and executive functioning network. Therefore, a poor network between the amygdala and anterior cingulate cortex hampers emotional regulation, resulting in affective disruption (Maletic et al. 2007).

The brain reacts to stress by activation of the HPA axis, wherein the hypothalamus responds by secreting CRF. CRF stimulates the synthesis and release of cortisol, which has an impact on general metabolism and behaviour through a series of mechanisms (Nestler et al. 2002). Cortisol regulates its own secretion via a negative feedback control mechanism by activation of the HPA axis (Nandam et al. 2020). The sustained high levels of cortisol damage hippocampal neurones and reduce the birth of new neurones (Nestler et al. 2002) Depleted levels of monoamines, hyperactivity of monoamine receptors, and decreased BDNF levels are the mechanisms underpinning depression (Nabeshima and Kim 2013) and biological and environmental factors, along with personal vulnerabilities, increase the sensitivity to the effects of the stressors.

9.8 TREATMENT AND CONTROL OF DEPRESSION

Depression, one of the leading causes of disability and morbidity, needs timely identification and treatment. It is vital to manage depression through various modalities. These modalities are shown in Figure 9.3.

9.8.1 Pharmacological Treatment

Antidepressant drugs are the first line of treatment used to manage depression. Antidepressant drugs are prescribed depending on the patient's symptoms, severity and duration of depression, and needs of the patient. The efficacy of the prescribed drugs depends on the response of the patient which in turn, depends on many factors like any existing comorbidity, duration of the disorder, chances of relapse, etc. These antidepressants aid in balancing levels of neurotransmitters and reducing symptoms like inflammation, thereby treating depression.

Antidepressants can be broadly classified into five classes: tricyclic antidepressants (TCAs), monoamine oxidase inhibitors (MAOIs), selective serotonin reuptake inhibitors (SSRIs), dual serotonin-norepinephrine reuptake inhibitors (SNRIs), and atypical antidepressants (Ogbru 2021).

Currently available antidepressants have been classified into two categories: classical and non-classical antidepressants. The classical antidepressants are MAOIs and TCAs. The non-classical antidepressants include SSRIs, SNRIs, serotonin receptor antagonism with serotonin reuptake inhibition (SARI), serotonin 5-HT_{1A}-autoreceptor partial agonism with serotonin reuptake inhibition (SPARI), serotonin-norepinephrine reuptake inhibition and serotonin receptor antagonism antidepressant with potent antipsychotic D2 receptor blockade/antagonism (SNRISA), norepinephrine reuptake inhibition with serotonin receptor antagonism (NRISA), noradrenergic α_2-receptor antagonism with specific serotonergic receptors-2 and-3 antagonism (NASSA), norepinephrine reuptake

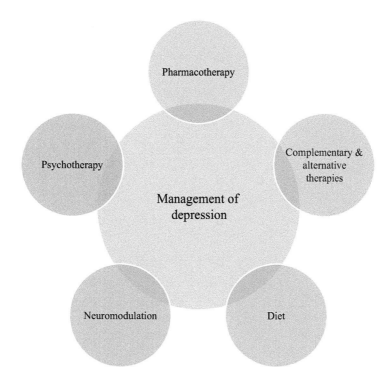

FIGURE 9.3   Management of depression.

inhibition (NRI), dual norepinephrine-dopamine reuptake inhibition (NDRI), atypical antipsychotics that exhibit weak D_2-receptor antagonism with potently strong $5\text{-}HT_{2A}$-receptor blockade and NMDA-glutamatergic ionoceptor antagonist/inverse agonist/partial agonist (Fasipe 2018).

MAOIs and TCAs are the classical antidepressant agents. MAOIs block the activity of the monoamine oxidase enzyme. TCAs work by blocking the reabsorption of neurotransmitters along with blocking certain receptors. They block norepinephrine and serotonin uptake pumps. TCAs and MAOIs are the initial classes of antidepressants, which are not a very preferable option anymore because of their high side effects compared to newer-generation antidepressant drugs (Ferguson 2001; Ogbru 2021).

SSRIs are preferred as a first-line treatment as they are relatively safe and tolerable (Kupfer 2005). They inhibit presynaptic serotonin uptake (Adams et al. 2008) via the inhibition of the serotonin reuptake transporter (Fasipe 2018). The non-classical antidepressant agents function by blocking one or more of the reuptake transporter pumps and/or receptors for the three monoaminergic neurotransmitters. For instance, SNRIs block the uptake of both serotonin and norepinephrine. At a low dose, they function as SSRIs and at a higher dose, they increase dopaminergic transmission as well. NDRIs inhibit the noradrenergic and dopaminergic reuptake transporter pump systems. SARIs exhibit the pharmacological property of a moderate-to-strong serotonin receptor(s) antagonism with weak serotonin reuptake transporter (SERT) inhibition (Fasipe 2018).

It has been reported that 40–60% of adults with moderate or severe depression notice improvement in their symptoms within 6–8 weeks, depending on the severity of illness, history of antidepressant responsiveness, and associated comorbidities (IQWiG 2020). Cipriani et al. (2018) in their systematic review and network meta-analysis found that out of 21 antidepressant drugs, escitalopram, mirtazapine, paroxetine, agomelatine, and sertraline had relatively higher response and lower dropout rates, indicating the efficacy and acceptability of these drugs in adults. Escitalopram was found to be superior in terms of response rate and acceptability by Andrade (2018). Amitriptyline, a

TCA, and fluoxetine, an SSRI, are considered as essential medicines for treating depressive symptoms (WHO 2021b). On the other hand, fluoxetine was found to be more effective in treating depression in children and adolescents (Cipriani et al. 2016). Anti-inflammatory drugs are also considered as a treatment option for depressive subjects, and it was found that corticosteroids were superior to other anti-inflammatory agents in terms of efficacy, non-steroidal anti-inflammatory drugs (NSAIDs) in terms of acceptability, and N-acetylcysteine in terms of remission response (Hang et al. 2021). Combining antidepressant drugs is a good option advocated if patients do not respond to monotherapy (Palaniyappan et al. 2009) or changing the medication to the same class or to a different class of drug if the response is inadequate (Adams et al. 2008). In fact, the choice of drug depends on the responsiveness of the patient and adverse effects of the drugs on the patient; therefore, healthcare providers should weigh the pros and cons of an antidepressant before administering.

Nevertheless, owing to the side effects, toxicity, and in some cases unresponsiveness of the patients towards pharmacological treatment, there arises a need to understand the suitability of non-pharmacological therapies in treating depression.

9.8.2 PSYCHOTHERAPY

Psychotherapies are alternatives or additional to pharmacological therapy for the treatment of depressive symptoms. Psychotherapy, also known as talk therapy, is a scientifically validated procedure to identify and change the emotions, cognitions, thoughts, and behaviour pattern of an individual for healthy wellbeing. It helps bring about a positive change in people suffering from mental illnesses like depression. Depression can be effectively treated with 6–8 sessions of 30 minutes of psychotherapy and is seen as an effective option for acute phase treatment of depression (Nieuwsma et al. 2012).

Various psychotherapy and basic elements underpinning the functioning of these therapies are given below (Parekh and Givon 2019; Cuijpers et al. 2008; Cuijpers et al. 2020a).

Cognitive Behaviour Therapy (CBT). This therapy deals with replacing harmful or ineffective thinking and behaviour with accurate thoughts and functional behaviour. It deals with practising new skills effective in solving current problems. It is aimed at the patient's dysfunctional beliefs and cognitive restructuring.

Interpersonal Therapy (IPT). As the name suggests, it deals with interpersonal issues that dysregulate the emotional balance, and therefore this therapy helps the individual to learn healthy ways to express emotions and communicate.

Psychodynamic Therapy (PT). It deals with creating self-awareness and changing old patterns which are troublesome for the individual. It is based on the principle that adverse childhood experiences influence mental wellbeing and behaviour and therefore by creating self-awareness one can manage the inappropriate repetitive thoughts or feelings.

Supportive Therapy (ST). This therapy aims at strengthening an individual's coping mechanism and building self-esteem and social and community functioning so that one becomes sufficiently equipped with his human resources to deal with the problems.

Life Review Therapy (LRT). This is reminiscence therapy and deals with difficult memories or unresolved issues from the past. It is aimed at systematically evaluating the life of the patient.

Problem Solving Therapy (PST). It deals with defining the personal problems of the patient, searching for alternatives to solve the problems, choosing the best alternative, implementing the solution, and finally evaluating to check the efficacy of the solution in resolving the problem.

Social Skill Training (SST). This therapy deals with developing skills like assertiveness that will help to build and retain social and personal relationships.

Behaviour Activation Therapy (BAT). It aims to change the way a person interacts with their environment. Pleasant activity scheduling is the core element of this therapy.

A host of research has concluded the effectiveness of psychotherapies in the treatment of depression. However, the severity of depression and acceptability of therapy by the patient determine the selection of the type of psychotherapy in combination with other treatment modalities. Psychotherapies for depression have been seen to improve the quality of life, mental functioning, social and work-related relationships, comfort level, and engagement in everyday activities (Kolovos et al. 2016). Psychotherapies for depression can be adopted for all age groups, however, psychotherapies for depression were observed to be most suitable for young adults compared to children and adolescents or middle-aged adults. However, the effect of psychotherapy was the same in middle-aged, old, and older-old adults (Cuijpers et al. 2020b).

Cuijpers et al. (2021) compared the efficacy, acceptability, and long-term outcome of psychotherapies and found that CBT, IPT, PST, BAT, and LRT were effective in depression compared to control treatments. Non-directive supportive counselling was less efficacious than other therapies. All therapies were acceptable in the treatment of acute depression and had a significant effect at a 12-month follow-up, with problem-solving therapy somewhat more effective than others. Barth et al. (2013) were also in agreement with the effectiveness of psychotherapies in depression and concluded that supportive counselling, SST, PST, CBT, BAT, and PT were more effective than control (waitlist). Treatment with CBT, IPT, and ST reduced depressive symptoms significantly compared to usual care (Health Quality Ontario 2017).

CBT as first-line treatment or as an adjunct to pharmacotherapy is effective in mild, moderate, severe, and chronic/recurrent depression and bipolar disorders (Gautam et al. 2020; Driessen and Hollon 2010). CBT not only decreases depressive scores but also improves mental health and quality of life (Reavell et al. 2018). It can maintain the reduced depression scores for a year as it can control automatic negative thoughts (Chiang et al. 2015). CBT was found to be more effective in female patients, patients with a higher rate of comorbid anxiety disorders (Braun et al. 2013), married subjects, patients who had low levels of pretreatment dysfunctional attitudes, unemployed individuals, had more antecedent life events, and prior exposure to antidepressants (Driessen and Hollon 2010). CBT along with fluoxetine, was found to be superior to placebo and monotherapies and in reducing suicidal thinking in adolescents (March et al. 2004). Similarly, IPT is also an established therapy for treating depression. IPT efficaciously treats acute depression and in combination with pharmacotherapy, it was more efficient than pharmacotherapy alone. The combination therapy was effective not only treating severe or chronic depression but also reduced the relapse rate (Cuijpers et al. 2011). IPT was found to be superior to treatment as usual, increased work ability, reduced work-related stress and problems with the maintenance of results in a three-month follow-up (Schramm et al. 2020). IPT was found to be more effective than other therapies, namely non-directive supportive treatment, BAT, PT, PST, and SST (Cuijpers et al. 2008). Psychodynamic therapy was found to be more effective in depression compared with other treatment modalities (Ribeiro et al. 2018; Steinert et al. 2017). Behaviour activation therapy can also be adopted as an option for treating depression as it was found to be more effective than humanistic therapy, medication, and treatment as usual and was on par with CBT and psychodynamic therapy (Uphoff et al. 2020). BAT was effective in older depressed patients as an individual format (Braun et al. 2013). BAT as group therapy in eight sessions for 1.5 hours reduced depressive symptoms in female subjects as these subjects needed attention, affection, and companionship (Ghanatghestani and Vaziri 2018). Similar results were found by Blanchet and Provencher (2020) where a change in depression, reinforcement, anxiety, social adjustment, and quality of life was seen along with maintenance of scores for four weeks. The effectiveness of PST in depression was concluded by Zhang et al. (2018). PST was seen to be effective not only reducing depression but also in reducing suicidal risks. It increased self-esteem and the level of assertiveness in participants, and the improvements were maintained at 12-month follow-up (Eskin et al. 2008). Conversely, it was also concluded that PST was more effective when used along with some other treatment modalities (Krause et al. 2021). SST can also be an effective tool in treating depression as it helps in effective communication and interaction, failing to which may lead to isolation and eventually depression. SST was found to be more effective in treating depression in university students, and a longer intervention programme led to a better outcome (Ndegwa 2021). SST had positive effects on self-esteem and improved interpersonal skills besides depression

(El Malky et al. 2016). LRT was found to be more effective in reducing depression compared to treatment as usual. LRT was also helpful in reducing anxiety and improving mental health (Korte et al. 2012), self-esteem, and obsessive reminiscence in older adults (Preschl et al. 2012). Group reminiscence therapy, which is more cost-effective than individual reminiscence therapy, was more effective in treating mild to moderate depression, increasing life satisfaction, psychological wellbeing, and decreasing loneliness (Liu et al. 2021).

To sum up, we can say that all the psychotherapies are effective in relieving depressive symptoms either alone or as combination therapy depending upon the right judgement of psychologists in identifying and understanding the needs of patients.

9.8.3 Neuromodulatory Therapy

Neuromodulation therapy is based on the principle that psychiatric illnesses like depression arise because of dysfunction in normal communication within a network of brain regions that regulate mood, thoughts, and behaviour and therefore targeting the altered neural structures with the use of neuromodulatory techniques which aid in treating depression. Neuromodulatory therapies become more significant in treatment-resistant depression and in high remission depression. They may be used as an alternative or adjunctive treatments for depression and can be categorised as non-invasive and invasive techniques. The non-invasive techniques include electroconvulsive therapy, transcranial magnetic stimulation, magnetic seizure therapy, and transcranial direct current stimulation, whereas invasive techniques are vagus nerve stimulation, deep brain stimulation, ablative surgery, and direct cortical stimulation.

9.8.3.1 Electroconvulsive Therapy (ECT)

It is a procedure where electric current is passed through the brain to induce generalised seizures. Depending on the target site of the brain, it is of three types, namely, bitemporal ECT, right unilateral ECT, and bifrontal ECT (Mutz et al. 2019; Holtzheimer and Mayberg 2012). ECT is a well-established method to treat major depression and treatment-resistant depression. It is relatively safer with a good remission rate and negligible relapse rate. However, it has certain side effects like headaches, vomiting, nausea, and cognitive withdrawal. Bilateral ECT is more efficient than unilateral ECT (Brunoni et al. 2010) but less cognitive withdrawal was seen in unilateral and ultra-brief pulse application (Li et al. 2020).

9.8.3.2 Transcranial Magnetic Stimulation (TMS)

In this therapy, electromagnetic fields are used to alter neural activity in a relatively focal superficial area of the brain. Repetitive TMS (rTMS) involves the delivery of pulses in a repetitive manner. A magnetic field is created on the surface of the scalp. High frequency stimulation of the left dorsolateral prefrontal cortex and low-frequency stimulation of the right dorsolateral prefrontal cortex help alleviate depressive symptoms. It has a good response and remission rate and a low relapse rate (Rizvi and Khan 2019; Mutz et al. 2019; Holtzheimer and Mayberg 2012).

9.8.3.3 Magnetic Seizure Therapy (MST)

In this therapy, magnetic fields are used to induce generalised seizures at the prefrontal cortex. Since it does not involve the delivery of non-focal and large total electric charge as in the case of ECT, which is probably responsible for cognitive withdrawal, MST has been reported to be relatively safer than ECT in terms of adverse cognitive effects. High-frequency stimulation was reported to have a better remission and response rate (Mutz et al. 2019; Daskalakis et al. 2020).

9.8.3.4 Transcranial Direct Current Stimulation

In this therapy, weak direct currents are applied to the brain through scalp electrodes that modulate cortical excitability according to current polarity (Brunoni et al. 2010). Mild temporary side effects like headache, redness of the skin, acute mood changes, and nausea are reported in the therapy (Thair et al. 2017).

9.8.3.5   Vagus Nerve Stimulation (VNS)

In this technique, an electrical pulse generator is surgically implanted under the skin of the chest and connected to the vagus nerve in the neck. The pulse stimulates the vagus nerve which regulates mood. Left cervical VNS is an approved therapy for treatment-resistant depression (Howland 2014). The target site is the left vagus nerve with the purpose to induce potentials on the ventroposterior complex and intralaminar areas through the thalamocortical pathways, thereby affecting the hypothalamus, amygdala, and other cortical structures (Bakar et al. 2019). It is effective in chronic and refractory depression.

9.8.3.6   Deep Brain Stimulation (DBS)

In this surgical technique, stimulatory electrodes are placed in specified brain areas which produce inhibition by depolarisation blockade and excitatory axonal response and alteration in the activity in neuronal circuits (Bakar et al. 2019). DBS is considered an effective option in treatment-resistant depression as it has shown a good response and remission rate and lower relapse rate. However, adverse effects like suicide and suicidal attempts have also been reported (Wu et al. 2021). Therefore, it is concluded that since it involves a greater risk, so it should be preferred as an option for severely ill patients (Holtzheimer and Mayberg 2012).

9.8.3.7   Ablative Surgery

It is a technique to treat neurological disorders by creating therapeutic lesions in the brain to break off maladaptive cerebral networks and to destroy abnormal tissues. The four main techniques used are radiofrequency thermoablation, stereotactic radiosurgery, laser interstitial thermal therapy, and high-intensity focused ultrasound thermal ablation. Three lesion procedures, namely anterior cingulotomy, subcaudate tractotomy, and limbic leucotomy, are performed in major depressive disorder (Franzini et al. 2019).

9.8.3.8   Direct Cortical Stimulation (DCT)

In this therapy, electrodes are surgically implanted directly onto the surface of the brain, and stimulation is controlled through an implanted pulse generator. It is a well-tolerated therapy; however, the risk of bleeding and infection due to surgery may be possible (Holtzheimer and Mayberg 2012).

The possible mechanism underlying the efficacy of these neuromodulatory therapies in treating major depression and treatment-resistant depression. may be restoration of hemisphere balance, increase in BDNF and neurogenesis, regulation of monoamines, and increase in cortical activity. However, owing to the adverse effects associated with some of the neuromodulatory techniques, there is a need to optimise the application of technique in terms of magnitude of dose or placement of electrodes, etc. in relation to the response of the patient. Still, research in this arena is ongoing to improve efficacy and safety.

9.8.4   Diet and Supplements

Nutrients in the form of diet or as a supplement can play a major role in treating and managing mental disorders like depression. Nutrients are essential for the normal regulation of brain functions, and a disbalance in optimum nutrients may hamper the various essential activities and lead to the onset of depression. For instance, various nutrients play a major role in the synthesis of neurotransmitters, which are essential to maintain brain integrity. The role of various nutrients in depression has been analysed (Businaro et al. 2021; Ekong and Iniodu 2021).

Protein. Deficiency of protein is directly associated with abnormal neurotransmitter levels as neurotransmitters are derivatives of amino acids. Deficiency of tryptophan and tyrosine can reduce the levels of serotonin and dopamine, respectively, which have been proven to have an essential role in the onset of depression. Thus, a diet rich in milk, cheese, meat, eggs, chicken, fish, beans, nuts, and wholegrains acts as a source of protein and amino acids essential for the synthesis of neurotransmitters.

Omega 3 fatty acids. Omega 3 fatty acids are constituents of the membrane of brain cells and are also observed to influence the production of neurotrophic factors involved in neurogenesis. α-linolenic acid found in soy, walnut, canola oil, and docasahexonoic and eicosapentaenoic acid found in seafood, eggs, and animal products are good sources of omega 3 fatty acids, and thus optimum incorporation in the diet may have a protective role in depression.

Vitamins. Beta-carotene, vitamins C, and E act as antioxidants. These can protect against oxidative stress known to play a role in the pathogenesis of depression. Similarly, vitamins D, K, and B-vitamins via their involvement in multiple mechanisms, affect synthesis and regulation of neurotransmitters, thereby influence depression.

Minerals. Minerals like sodium, potassium, magnesium, calcium, phosphorus, iron, zinc, cobalt, and chromium have important functions in neural transmission, neuromuscular regulation, synaptic transmission, and antioxidation.

Probiotics may also play a role in mitigating depressive symptoms, possibly because of their role in reducing production of inflammatory chemicals and their modulatory effect on tryptophan, a precursor of serotonin (Liu et al. 2019; Noonan et al. 2020). Daily supplementation of probiotics for a period of 4–8 weeks was found to be more effective in reducing the severity of symptoms in depressive subjects with safety and good acceptability (Wallace and Milev 2021). However, more clinical studies are required to validate the potentiality of probiotics and other dietary supplements.

It has been seen that a high-quality diet is related with a lower risk of depression. Diet and dietary supplementation help the brain to function properly and also provide a neuroprotective effect. Thus, diet can both treat as well as prevent the depressive state (Owens et al. 2020). An unhealthy Western diet was seen to be associated with an increased prevalence of depression. and healthy diets like the Japanese and Mediterranean diets are inversely associated with depression (Lang et al. 2015; Khanna et al. 2019). The intake of less nutrient-dense food and unhealthy food rich in sodium, sugar, and fat results in poor brain development and increases the likelihood of depression. The trend of eating unhealthy and junk food was much observed in adolescents (Samuelson 2017). Intervention studies with a healthy diet (rich in fruit, vegetables, wholegrain cereals, protein, unsweetened dairy, fish, nuts and seeds, olive oil, spices like turmeric and cinnamon) showed lower depressive symptoms compared to the controls (Francis et al. 2019).

9.8.5 COMPLEMENTARY AND ALTERNATIVE MEDICINE (CAM) THERAPIES

According to National Center for Complementary and Alternative Medicine (NCCAM), Complementary and Alternative Medicine consists of 'a group of diverse medical and healthcare systems, practices, and products that are not presently considered to be part of conventional medicine' (Fan 2005). They are not part of standard healthcare but owing to negligible side effects and cost effectiveness they are preferred treatment options by many patients.

The most frequently used CAM therapies in depression are herbal medicine (borage, chamomile, lavender), followed by prayer therapy (Ashraf et al. 2021). The other commonly used methods are mindfulness, basal body awareness, massage/tactile stimulation, acupuncture, yoga, music, and light therapy (Wemrell et al. 2020). CAM therapies are used more than conventional therapies by severely depressed people (Kessler et al. 2001). The main CAM therapies used in depression are discussed below.

9.8.5.1 Herbs and Herbal Medicine

Various herbs have been identified as potent agents for ameliorating mild to moderate depressive symptoms. Ginseng, a Chinese herb, is used to improve mood. The compound 20(S)-protopanaxadiol isolated from ginseng showed antidepressant property. Similarly, *Paeonia lactiflora*, *Pall* (peony), *Albizia julibrissin* (mimosa or silk tree), *Epimedium* (barrenwort), *Perilla frutescens* (perilla), *Fuzi*, *Rhodiola rosea* (rose root), *Polygala tenuifolia* (milk wort), and saffron have been reported to have antidepressive properties. Xia Yao San, a decoction made from eight herbs, was stated to possess antidepressant properties in Chinese medicine (Liu et al. 2015). Lavender, passion flower, black cohosh, chamomile, chasteberry, and saffron showed results in mitigating depression in cancer patients (Yeung et al. 2018). Green tea, saffron, borage, St John's wort, kava (*Piper methysticum)*, roseroot, lavender, and sacred lotus (*Nelumbo nucifera*) fruit were reported to be effective against depression (Lee and Bae 2017). Setorki (2020) concluded that St John's wort, saffron, curcumin, chamomile, lavender, and rose root improved mild to moderate depression. The phytochemicals present in these herbs have antidepressant effects. For instance, carvacrol present in oregano and thyme, curcumin present in turmeric, L-theanine in green tea, proanthocyanidin present in apple, cocoa bean, grape, and tea, quercetin in some fruit and vegetables, resveratrol in grapes and red wine possess antidepressant effect (Lee and Bae 2017). Human clinical studies showed that borage (*Echium amoenum*) and roseroot were effective in reducing depression symptoms compared to placebo at four weeks of supplementation. Lavender was seen to boost the efficacy of antidepressant drugs. Lavender, along with antidepressant drugs, was more effective in reducing depression compared to the antidepressant alone. Saffron significantly reduced depression compared to placebo and when compared with antidepressant drugs, it showed equivalent therapeutic response. St John's wort was also seen to be significantly more effective than placebo and comparable to SSRIs effect (Sarris et al. 2011). The bioactive components present in these herbs help to ameliorate depression symptoms by suppressing the reuptake of neurotransmitters in the brain and levels of corticosterone, inhibiting monoamine oxidase, regulating BDNF and HPA axis, thereby promoting neuronal cell survival and differentiation (Lee and Bae 2017; Liu et al. 2015).

These herbs are proven to be less toxic and have negligible adverse effects compared to antidepressant drugs. However, more human clinical research is warranted to conclude their antidepressant effects.

9.8.5.2 Exercise

Exercise has a therapeutic effect on depression in all age groups (mostly 18–65 years old), as a single therapy, or an adjuvant therapy, or a combination therapy. The benefits of exercise therapy are comparable to traditional treatments for depression. Moderate-intensity exercise is enough to reduce depressive symptoms, but higher-dose exercise is better for overall functioning. Exercise therapy has become more popular because of its benefits to the cardiovascular system, emotional state, and systemic functions. Aerobic exercise/mind-body exercise (3–5 sessions per week with moderate intensity lasting for 4–16 weeks) is recommended. Individualised protocols in the form of group exercise with supervision are effective at increasing adherence to treatment (Xie et al. 2021). Exercise combined with standard treatment yielded significant antidepressant effects compared to standard treatment alone (Lee et al. 2021). For instance, exercise in conjunction with antidepressants was significantly more effective in lowering depression than antidepressant alone (Tasci et al. 2019).

Aerobic exercise at a dose consistent with public health recommendation (17.5 kcal/kg/week) was effective in treating mild to moderate major depressive disorder (Dunn et al. 2005). Supervised weightlifting had an antidepressant effect, and the effect persisted even after direct supervision was withdrawn, indicating no relapse rate in the elderly with severe depression even after the removal of supervised exercise (Singh et al. 2001). Resistance exercise training was also seen to be effective in decreasing depressive symptoms in adults (Gordon et al. 2018). Aerobic exercise and stretching decreased depression and improved self-efficacy and episodic memory, implying their role in bringing positive psychological and cognitive changes in depressed subjects (Foley et al. 2008). Exercise is suggested to increase the availability of neurotransmitters, stimulate the growth of nerve cells, increase BDNF, decrease pro-inflammatory cytokines, and modulate HPA axis, thus regulating neurogenesis, neuroprotection, and synaptic plasticity (Al-Qahatani et al. 2018; Matta Mello Portugal et al. 2013).

9.8.5.3   Mindfulness Therapy

Mindfulness is a therapy that helps to reach a mental state characterised by non-judgemental awareness of present moment experiences. It teaches to pay attention to one's thoughts and feelings without being judgemental. The two mindfulness approaches are mindfulness-based stress reduction (MBSR) and mindfulness-based cognitive therapy (MBCT).

The MBSR technique, initially developed to manage stress, is found to be more effective in treating depression (Hofmann et al. 2010; Chi et al. 2018). It is a formal training involving meditation, yoga, stretching, and body scan. Mindfulness encourages openness, curiosity, and acceptance. The two components of mindfulness are self-regulation of attention and orientation towards the present moment, effective in reducing depression (Hofmann et al. 2010). MBCT, an extension of MBSR, is a combination of mindfulness training along with cognitive therapy to reduce depression and its recurrence. In this therapy, internal awareness is improved to disengage from maladaptive patterns of repetitive negative thinking (Hofmann and Gomez 2017). These techniques are usually an eight-week programme where the practitioner observes sensations, emotions, and thoughts with a non-reactive attitude. A drop in depression score with maintenance up to two months of follow-up was observed after eight weeks of mindfulness therapy. Self-compassion, non-reactivity, and non-judging were seen to be improved (Takahashi et al. 2019). MBCT significantly reduced depression severity and response rate and increased the remission rate when compared to a Health Enhancement Programme (Physical fitness + music therapy + nutrition education), concluding the effectiveness of MBCT as an adjunctive treatment with medication in treatment-resistant depression (Eisendrath et al. 2016).

Mindfulness therapy is presumed to indirectly influence depression by regulating emotions. It has an impact on worry, rumination, reappraisal, and suppression. The therapy reduces worry, rumination, and suppression and is positively associated with reappraisal (Parmentier at al 2019). The antidepressive effect is seen to be mediated by self-compassion, cognitive reactivity, aversion, positive affect, rumination, and worry (Maddock and Blair 2021). Alteration in these dimensions plays a role in reducing depression and its relapsing risk (van der Velden et al. 2015).

Mindfulness techniques can be effectively used to exercise attentiveness. The therapy makes an individual attentive enough to recognise and quickly intervene with the negative feelings and worries that may otherwise lead to depression.

A few more therapies like light therapy, prayer therapy, aromatherapy, acupuncture, and massage therapy have been investigated for their usefulness in ameliorating symptoms of depression. Light therapy is used to treat seasonal affective disorders (Reeves et al. 2012), however it is found to be equally effective in non-seasonal disorders as well as an adjunctive or alternative therapy. Bright light therapy is effective but may have minor adverse effects like headaches, eyestrain, nausea (Maruani and Geoffroy 2019; Lam et al. 2016). Light therapy is found to stabilise mood, improve sleep quality, and stabilise circadian rhythm. It is found to be more effective when applied for less than one hour in a day (Tao et al. 2020). It was concluded that bright light therapy as monotherapy or in combination with an antidepressant drug was effective in non-seasonal depression compared to a placebo (Lam et al. 2016). Prayer therapy was also found to be more effective in lowering depression score and increasing the level of optimism compared to the control group. The prayer intervention group was able to maintain the lower depression score during one month follow-up (Boelens et al. 2009) and until one year follow-up (Boelens et al. 2012). Aromatherapy (inhalation and massage aromatherapy) was found to be more effective in alleviating depressive symptoms (Sanchez-Vidana et al. 2017). Aromatherapy with lavender may have the potential to improve mild to moderate depression (Jafari-Koulaee et al. 2020). Acupuncture was also seen to be effective in reducing the severity of depression (Armour et al. 2019). Both electro-acupuncture and manual acupuncture were effective in improving depression when given alongside SSRIs as compared to SSRIs or acupuncture monotherapy (Zhichao et al. 2021). Massage therapy was found to alleviate depressive symptoms (Hou et al. 2010). It was seen that the intervention group receiving a massage which was a one-hour special gentle touch technique for four weeks was effective in reducing depression symptoms and general somatic symptoms (Arnold et al. 2020).

These above-discussed complementary therapies might be more useful when used as adjunctive therapy. However, conclusive research is warranted in this arena.

9.9 PREVENTIVE MEASURES

Depression, a mental illness, needs to be prevented first-hand in order to reduce the physical, psychological, and financial burden associated with it, affecting not only the individual and their caregiver, but also society as a whole. The primary key to prevention is awareness. Each person should be aware of the factors affecting them adversely and know how to handle the stressors. This becomes especially important for individuals who are genetically susceptible to depression or are in close vicinity with those environmental factors that may increase the severity. Such individuals need to be identified and treated at an early stage, before the depression worsens. Therefore, the following tips are suggested to prevent depression:

1. Early identification of individuals with an increased susceptibility to depression
2. Creating a conducive home and school environment, especially for children
3. Making a favourable work environment for employees
4. Promoting any kind of physical exercise as per the suitability for various age groups
5. Building strong social relationships and cutting on the time spent on social media.
6. Preparing a feasible work schedule and avoiding unrealistic goals
7. Getting plenty of sleep
8. Enjoying each and every moment and avoiding nagging
9. Staying away from toxic situations and people
10. Eating organic and balanced food comprising of all the nutrients
11. Following a healthy lifestyle. Avoiding smoking, alcohol, and drug abuse
12. Conducting an awareness education programme as part of the curriculum in schools and universities
13. Cultivating hobbies and engaging in pleasurable activities
14. Connecting with nature
15. Expressing oneself through talking with close ones and not dwelling on stress alone
16. Seeking professional help at the very initial stage to prevent the exacerbation of the problem

9.10 CONCLUSION

Depression, a mental illness, is increasing at an alarming rate throughout the world. It can affect any age and any gender, with females more susceptible than males. Depression is a debilitating mental condition affecting the daily lives of the subject. The person loses interest and pleasure in activities and becomes surrounded by feelings of helplessness, low self-worth, and guilt. It has been seen that genetics, environment, and personal vulnerabilities interplay to increase the risk of depression induction. These stressors, if not handled properly, through a series of biological mechanisms in the brain and CNS, lead to the generation of depression, which, if not treated on time, can have fatal manifestations. With the advancement in the understanding of depression and its causal factors, methods of treatment such as pharmacological, psychotherapies, neuromodulatory techniques, and CAM therapies have been recognised. These therapies are proven to be efficacious in treating depression. Depending on the subject's responsiveness, tolerability, acceptability, duration, and severity of illness, the type of therapy may be chosen by health practitioners. Although depression is a crippling disorder affecting the normal functioning of a person, it can be treated effectively and can be prevented by following simple measures in daily life.

REFERENCES

Abbasi, H. 2021. The effect of climate change on depression in urban areas of western Iran. *BMC Research Notes* 14: 155. https://doi.org/10.1186/s13104-021-05565-0

Abela, J.R.Z., and B.L. Hankin. 2008. Cognitive vulnerability to depression in children and adolescents: A developmental psychopathology perspective. In *Handbook of Depression in Children and Adolescents*, edited by J.R.Z. Abela and B.L. Hankin. The Guilford Press, New York, 35–78.

Adams, S.M., K.E. Miller, and R.G. Zylstra. 2008. Pharmacological management of adult depression. *American Family Physician* 77, no. 6: 785–792.

Alhassan, A.A., E.M. Alqadhib, N.W. Taha, R.A. Alahmari, M. Salam, and A.F. Almutairi. 2018. The relationship between addiction to smartphone usage and depression among adults: A cross sectional study. *BMC Psychiatry* 18, no. 1: 148. https://doi.org/10.1186/s12888-018-1745-4

Al-Qahatani, A.M., M.A.K. Shaikh, and I.A. Shaikh. 2018. Exercise as a treatment modality for depression: A narrative review. *Alexandria Journal of Medicine* 54: 429–435.

Andrade, C. 2018. Relative efficacy and acceptability of antidepressant drugs in adults with major depressive disorder: Commentary on a network meta-analysis. *Journal of Clinical Psychiatry* 79, no. 2: 18f12254. https://doi.org/10.4088/JCP.18f12254

Andriola, E., M.D. Trani, A. Grimaldi, and R. Donfrancesco. 2011. The relationship between personality and depression in expectant parents. *Depression Research and Treatment* 2011: 356428. https://doi.org/10.1155/2011/356428

APA. 2013. *Diagnostic and Statistical Manual of Mental Disorders: 5th Edn.* APA, Washington, DC.

Armour, M., C.A. Smith, L.Q. Wang, et al. 2019. Acupuncture for depression: A systematic review and meta-analysis. *Journal of Clinical Medicine* 8, no. 8: 1140. https://doi.org/10.3390/jcm8081140

Arnold, M.M., B. Muller-Oerlinghausen, N. Hemrich, and D. Bonsch. 2020. Effects of psychoactive massage in outpatients with depressive disorders: A randomized controlled mixed-methods study. *Brain Sciences* 10, no. 10: 676. https://doi.org/10.3390/brainsci10100676

Arvind, B.A., G. Gururaj, and S. Loganathan. 2019. Prevalence and socioeconomic impact of depressive disorders in India: Multisite population-based cross-sectional study. *BMJ Open* 9, no. 6: e027250. https://doi.org/10.1136/bmjopen-2018-027250

Ashraf, H., A. Salehi, M. Sousani, and M.H. Sharifi. 2021. Use of complementary alternative medicine and the associated factors among patients with depression. *Evidence-Based Complementary and Alternative Medicine: eCAM* 2021: 6626394. https://doi.org/10.1155/2021/6626394

Bagheri Hosseinabadi, M., N. Khanjani, M.H. Ebrahimi, B. Haji, and M. Abdolahfard. 2019. The effect of chronic exposure to extremely low-frequency electromagnetic fields on sleep quality, stress, depression and anxiety. *Electromagnetic Biology and Medicine* 38, no. 1: 96–101.

Bakar, B., C. Cetin, J. Oppong, and A.M. Erdogan. 2019. Current ablation type surgical treatment modalities in treatment-resistant major depression: Review of the recent major surgical series. *Journal of Basic and Clinical Health Sciences* 3, no. 1: 1–8.

Balsamo, M., C. Imperatori, M.R. Sergi, et al. 2013. Cognitive vulnerabilities and depression in young adults: An ROC curves analysis. *Depression Research and Treatment* 2013: 407602. https://doi.org/10.1155/2013/407602

Barth, J., T. Munder, H. Gerger, et al. 2013. Comparative efficacy of seven psychotherapeutic interventions for patients with depression: A network meta-analysis. *Plos Medicine* 10, no. 5: e1001454. https://doi.org/10.1371/journal.pmed.1001454

Beattie, G.S. 2005. Social causes of depression. http://www.personalityresearch.org/papers/beattie.html (accessed 28 May 2022).

Beutel, M.E., C. Junger, E.M. Klein, et al. 2016. Noise annoyance is associated with depression and anxiety in the general population- The contribution of aircraft noise. *PLoS One* 11, no. 5: e0155357. https://doi.org/10.1371/journal.pone.0155357

Blanchet, V., and M.D. Provencher. 2020. Effectiveness of behavioral activation for the treatment of severe depression in clinical settings. *Sante Mentale au Quebec* 45, no. 1: 11–30.

Boelens, P.A., R.R. Reeves, W.H. Replogle, and H.G. Koenig. 2009. A randomized trial of the effect of prayer on depression and anxiety. *International Journal of Psychiatry in Medicine* 39, no. 4: 377–392.

Boelens, P.A., R.R. Reeves, W.H. Replogle, and H.G. Koenig. 2012. The effect of prayer on depression and anxiety: Maintenance of positive influence one year after prayer intervention. *International Journal of Psychiatry in Medicine* 43, no. 1: 85–98.

Boers, E., M.H. Afzali, N. Newton, and P. Conrod. 2019. Association of screen time and depression in adolescence. *JAMA Pediatrics* 173, no. 9: 853–859. https://doi.org/10.1001/jamapediatrics.2019.1759

Boyce, P., G. Parker, B. Barnett, M. Cooney, and F. Smith. 1991. Personality as a vulnerability factor to depression. *British Journal of Psychiatry* 159: 106–114.

Braun, S.R., B. Gregor, and U.S. Tran. 2013. Comparing bona fide psychotherapies of depression in adults with two meta-analytical approaches. *PLoS One* 8, no. 6: e68135. https://doi.org/10.1371/journal.pone.0068135

Brunoni, A.R., C.T. Teng, C. Correa, et al. 2010. Neuromodulation approaches for the treatment of major depression: Challenges and recommendations from a working group meeting. *Arquivos de Neuro-Psiquiatria* 68, no. 3: 433–451. https://doi.org/10.1590/s0004-282x2010000300021

Businaro, R., D. Vauzour, J. Sarris, et al. 2021. Therapeutic opportunities for food supplements in neurodegenerative disease and depression. *Frontiers in Nutrition* 8: 669846. https://doi.org/10.3389/fnut.2021.669846

CDC. 2021. About mental health. National Center for Chronic Disease Prevention and Health Promotion. Division of Population Health. https://www.cdc.gov/mentalhealth/learn/index.htm (accessed 25 July 2022).

Chi, X., A. Bo, T. Liu, P. Zhang, and I. Chi. 2018. Effects of mindfulness-based stress reduction on depression in adolescents and young adults: A systematic review and meta-analysis. *Frontiers in Psychology* 9: 1034. https://doi.org/10.3389/fpsyg.2018.01034

Chiang, K.J., T.H. Chen, H.T. Hsieh, J.C. Tsai, K.L. Ou, and K.R. Chou. 2015. One-year follow-up of the effectiveness of cognitive behavioral group therapy for patients' depression: A randomized, single-blinded, controlled study. *Scientific World Journal* 2015: 373149. https://doi.org/10.1155/2015/373149

Cianconi, P., S. Betro, and L. Janiri. 2020. The impact of climate change on mental health: A systematic descriptive review. *Frontiers in Psychiatry* 11: 74. https://doi.org/10.3389/fpsyt.2020.00074

Cipriani, A., T.A. Furukawa, G. Salanti, et al. 2018. Comparative efficacy and acceptability of 21 antidepressant drugs for the acute treatment of adults with major depressive disorder: A systematic review and network meta-analysis. *Lancet* 391, no. 10128: 1357–1366. https://doi.org/10.1016/S0140-6736(17)32802-7

Cipriani, A., X. Zhou, G. Del Giovane, et al. 2016. Comparative efficacy and tolerability of antidepressants for major depressive disorder in children and adolescents: A network meta-analysis. *Lancet* 388, no. 10047: 881–890. https://doi.org/10.1016/S0140-6736(16)30385-3

Cordova-Palomera, A., C. Tornador, C. Falcon, et al. 2016. Environmental factors linked to depression vulnerability are associated with altered cerebellar resting-state synchronization. *Scientific Reports* 6: 37384. https://doi.org/10.1038/srep37384

COVID-19 Mental Disorders Collaborators. 2021. Global prevalence and burden of depressive and anxiety disorders in 204 countries and territories in 2020 due to the COVID-19 pandemic. *The Lancet* 398, no. 10312: 1700–1712. https://doi.org/10.016/S0140-6736(21)02143-7.

Cuijpers, P., A. van Straten, G. Andersson, and P. van Oppen. 2008. Psychotherapy for depression in adults: A meta-analysis of comparative outcome studies. *Journal of Consulting and Clinical Psychology* 76, no. 6: 909–922. https://doi.org/10.1037/a0013075

Cuijpers, P., A.S. Geraedts, P. van Oppen, G. Andersson, J.C. Markowitz, and A. van Straten. 2011. Interpersonal psychotherapy for depression: A meta-analysis. *American Journal of Psychiatry* 168, no. 6: 581–592. https://doi.org/10.1176/appi.ajp.2010.10101411

Cuijpers, P., E. Karyotaki, L. de Wit, and D.D. Ebert. 2020a. The effects of fifteen evidence-supported therapies for adult depression: A meta-analytic review. *Psychotherapy Research* 30, no. 3: 279–293. https://doi.org/10.1080/10503307.2019.1649732

Cuijpers, P., E. Karyotaki, D. Eckshtain, et al. 2020b. Psychotherapy for depression across different age groups: A systematic review and meta-analysis. *JAMA Psychiatry* 77, no. 7: 694–702. https://doi.org/10.1001/jamapsychiatry.2020.0164

Cuijpers, P., S. Quero, H. Noma, et al. 2021. Psychotherapies for depression: A network meta-analysis covering efficacy, acceptability and long-term outcomes of all main treatment types. *World Psychiatry* 20, no. 2: 283–293. https://doi.org/10.1002/wps.20860

Cutrona, C.E., G. Wallace, and K.A. Wesner. 2006. Neighborhood characteristics and depression: An examination of stress processes. *Current Directions in Psychological Science* 15, no. 4: 188–192. https://doi.org/10.1111/j.1467-8721.2006.00433.x

Dagnino, P., M.J. Ugarte, F. Morales, S. González, D. Saralegui, and J.C. Ehrenthal. 2020. Risk factors for adult depression: Adverse childhood experiences and personality functioning. *Frontiers in Psychology* 11: 594698. https://doi.org/10.3389/fpsyg.2020.594698

Daskalakis, Z.J., J. Dimitrova, S.M. McClintock, et al. 2020. Magnetic seizure therapy (MST) for major depressive disorder. *Neuropsychopharmacology* 45, no. 2: 276–282. https://doi.org/10.1038/s41386-019-0515-4

Driessen, E., and S.D. Hollon. 2010. Cognitive behavioral therapy for mood disorders: Efficacy, moderators and mediators. *Psychiatric Clinics of North America* 33, no. 3: 537–555. https://doi.org/10.1016/j.psc.2010.04.005

Dunn, A.L., M.H. Trivedi, J.B. Kampert, C.G. Clark, and H.O. Chambliss. 2005. Exercise treatment for depression: Efficacy and dose response. *American Journal of Preventive Medicine* 28, no. 1: 1–8. https://doi.org/10.1016/j.amepre.2004.09.003

Echezarraga, A., L. Fernández-Gonzalez, and E. Calvete. 2021. The role of temperament traits as predictors of depressive symptoms and resilience in adolescents. *Journal of Research in Personality* 95: e104155.

Eisendrath, S.J., E. Gillung, K.L. Delucchi, et al. 2016. A randomized controlled trial of mindfulness-based cognitive therapy for treatment-resistant depression. *Psychotherapy and Psychosomatics* 85, no. 2: 99–110.

Ekong, M.B., and C.F. Iniodu. 2021. Nutritional therapy can reduce the burden of depression management in low income countries: A review. *IBRO Neuroscience Reports* 11: 15–28.

El Malky, M.I., M.M. Attia, and F.H. Alam. 2016. The effectiveness of social skill training on depressive symptoms, self-esteem and interpersonal difficulties among schizophrenic patients. *International Journal of Advanced Nursing Studies* 5, no. 1: 43–50.

Elovainio, M., M. Jokela, T. Rosenstrom, et al. 2015. Temperament and depressive symptoms: What is the direction of the association? *Journal of Affective Disorders* 170: 203–212.

Enns, M.W., and B.J. Cox. 1997. Personality dimensions and depression: Review and commentary. *Canadian Journal of Psychiatry* 42, no. 3: 274–284.

Eskin, M., K. Ertekin, and H. Demir. 2008. Efficacy of a problem-solving therapy for depression and suicide potential in adolescents and young adults. *Cognitive Therapy and Research* 32: 227–245. https://doi.org/10.1007/s10608-007-9172-8

Fan, K.W. 2005. National center for complementary and alternative medicine website. *Journal of the Medical Library Association* 93, no. 3: 410–412.

Fasipe, O.J. 2018. Neuropharmacological classification of antidepressant agents based on their mechanisms of action. *Archives of Medicine and Health Sciences* 6: 81–94.

Fava, M., and K.S. Kendler. 2000. Major depressive disorder. *Neuron* 28, no. 2: 335–341.

Ferguson, J.M. 2001. SSRI antidepressant medications: Adverse effects and tolerability. *Primary Care Companion to the Journal of Clinical Psychiatry* 3, no. 1: 22–27. https://doi.org/10.4088/pcc.v03n0105

Fernandez-Pujals, A.M., M.J. Adams, and P. Thomson. 2015. Epidemiology and heritability of major depressive disorder, stratified by age of onset, sex, and illness course in generation Scotland: Scottish Family Health Study (GS:SFHS). *PLoS One* 10, no. 11: e0142197. https://doi.org/10.1371/journal.pone.0142197

Ferrari, A.J., F.J. Charlson, R.E. Norman, et al. 2013. Burden of depressive disorders by country, sex, age, and year: Findings from the global burden of disease study 2010. *Plos Medicine* 10, no. 11: e1001547.

Foley, L., H. Prapavessis, E.A. Osuch, J.A. De Pace, B.A. Murphy, and N.J. Podolinsky. 2008. An examination of potential mechanisms for exercise as a treatment for depression: A pilot study. *Mental Health and Physical Activity* 1, no. 2: 69–73.

Francis, H.M., R.J. Stevenson, J.R. Chambers, D. Gupta, B. Newey, and C.K. Lim. 2019. A brief diet intervention can reduce symptoms of depression in young adults - A randomised controlled trial. *PLoS One* 14, no. 10: e0222768. https://doi.org/10.1371/journal.pone.0222768

Franzini, A., S. Moosa, D. Servello, et al. 2019. Ablative brain surgery: An overview. *International Journal of Hyperthermia* 36, no. 2: 64–80. https://doi.org/10.1080/02656736.2019.1616833

Frewen, P.A., J. Brinker, R.A. Martin, and D.J.A. Dozois. 2008. Humor styles and personality-vulnerability to depression. *Humor – International Journal of Humor Research* 21, no. 2: 179–195.

Gautam, M., A. Tripathi, D. Deshmukh, and M. Gaur. 2020. Cognitive behavioral therapy for depression. *Indian Journal of Psychiatry* 62, no. S2: S223–S229. https://doi.org/10.4103/psychiatry.IndianJPsychiatry_772_19

Ghanatghestani, L.M., and S. Vaziri. 2018. Effectiveness of behavioral activation therapy on depression and social self-efficacy of depressed woman in Yazd. *Journal of Advanced Pharmacy Education and Research* 8, no. S2: 164–167.

Gordon, B.R., C.P. McDowell, M. Hallgren, J.D. Meyer, M. Lyons, and M.P. Herring. 2018. Association of efficacy of resistance exercise training with depressive symptoms: Meta-analysis and meta-regression analysis of randomized clinical trials. *JAMA Psychiatry* 75, no. 6: 566–576.

Gutierrez-Rojas, L., A. Porras-Segovia, H. Dunne, N. Andrade-Gonzalez, and J.A. Cervilla. 2020. Prevalence and correlates of major depressive disorder: A systematic review. *Brazilian Journal of Psychiatry* 42: 657–672.

Haddadi, P., and M.A. Beshrat. 2010. Resilience vulnerability and mental health. *Procedia – Social and Behavioral Sciences* 5: 639–642.

Hang, X., Y. Zhang, J. Li, et al. 2021. Comparative efficacy and acceptability of anti-inflammatory agents on major depressive disorder: A network meta-analysis. *Frontiers in Pharmacology* 12: 691200. https://doi.org/10.3389/fphar.2021.691200

Hankin, B.L., C. Oppenheimer, J. Jenness, A. Barrocas, B.G. Shapero, and J. Goldband. 2009. Developmental origins of cognitive vulnerabilities to depression: Review of processes contributing to stability and change across time. *Journal of Clinical Psychology* 65, no. 12: 1327–1338. https://doi.org/10.1002/jclp.20625

Health Quality Ontario. 2017. Psychotherapy for major depressive disorder and generalized anxiety disorder: A health technology assessment. *Ontario Health Technology Assessment Series* 17, no. 15: 1–167.

Himmerich, H., O. Patsalos, N. Lichtblau, M.A.A. Ibrahim, and B. Dalton. 2019. Cytokine research in depression: Principles, challenges, and open questions. *Frontiers in Psychiatry* 10: 30. https://doi.org/10.3389/fpsyt.2019.00030

Hofmann, S.G., and A.F. Gomez. 2017. Mindfulness-based interventions for anxiety and depression. *Psychiatric Clinics of North America* 40, no. 4: 739–749.

Hofmann, S.G., A.T. Sawyer, A.A. Witt, and D. Oh. 2010. The effect of mindfulness-based therapy on anxiety and depression: A meta-analytic review. *Journal of Consulting and Clinical Psychology* 78, no. 2: 169–183.

Holmans, P., M.M. Weissman, G.S. Zubenko, et al. 2007. Genetics of recurrent early-onset major depression (GenRED): Final genome scan report. *American Journal of Psychiatry* 164: 248–258.

Holtzheimer, P.E., and H.S. Mayberg. 2012. Neuromodulation for treatment-resistant depression. *F1000 Medicine Reports* 4: 22. https://doi.org/10.3410/M4-22

Hou, W.H., P.T. Chiang, T.Y. Hsu, S.Y. Chiu, and Y.C. Yen. 2010. Treatment effects of massage therapy in depressed people: A meta-analysis. *Journal of Clinical Psychiatry* 71, no. 7: 894–901.

Howland, R.H. 2014. Vagus nerve stimulation. *Current Behavioral Neuroscience Reports* 1, no. 2: 64–73. https://doi.org/10.1007/s40473-014-0010-5

Huh, H.J., S.Y. Kim, J.J. Yu, and J.H. Chae. 2014. Childhood trauma and adult interpersonal relationship problems in patients with depression and anxiety disorders. *Annals of General Psychiatry* 13: 26. https://doi.org/10.1186/s12991-014-0026-y

Ioannou, M., A.P. Kassianos, and M. Symeou. 2019. Coping with depressive symptoms in young adults: Perceived social support protects against depressive symptoms only under moderate levels of stress. *Frontiers in Psychology* 9: 2780. https://doi.org/10.3389/fpsyg.2018.02780

IQWiG. 2020. *Depression: How Effective are Antidepressants?* Institute for Quality and Efficiency in Health Care (IQWiG); 2006 [Updated 18 June 2020]. https://www.ncbi.nlm.nih.gov/books/NBK279282/?report=reader

Ivano, D.M. 2005. Marital problems: Do they cause depression? http://www.personalityresearch.org/papers/beattie.html (accessed 28 May 2022).

Jafari-Koulaee, A., F. Elyasi, Z. Taraghi, E.S. Ilali, and M. Moosazadeh. 2020. A systematic review of the effects of aromatherapy with lavender essential oil on depression. *Central Asian Journal of Global Health* 9, no. 1. https://doi.org/10.5195/cajgh.2020.442

Jylha, P., O. Mantere, T. Melartin, et al. 2011. Differences in temperament and character dimensions in patients with bipolar I or II or major depressive disorder and general population subjects. *Psychological Medicine* 41, no. 8: 1579–1591.

Kendler, K.S., M. Gatz, C.O. Gardner, and N.L. Pedersen. 2006. A Swedish national twin study of lifetime major depression. *American Journal of Psychiatry* 163, no. 1: 109–114.

Kessler, R.C., J. Soukup, R.B. Davis, et al. 2001. The use of complementary and alternative therapies to treat anxiety and depression in the United States. *American Journal of Psychiatry* 158, no. 2: 289–294. https://doi.org/10.1176/appi.ajp.158.2.289

Khanna, P., V.K. Chattu, and B.T. Aeri. 2019. Nutritional aspects of depression in adolescents – A systematic review. *International Journal of Preventive Medicine* 10: 42. https://doi.org/10.4103/ijpvm.IJPVM_400_18

Klein, D.N., R. Kotov, and S.J. Bufferd. 2011. Personality and depression: Explanatory models and review of the evidence. *Annual Review of Clinical Psychology* 7: 269–295. https://doi.org/10.1146/annurev-clinpsy-032210-104540

Kolovos, S., A. Kleiboer, and P. Cuijpers. 2016. Effect of psychotherapy for depression on quality of life: Meta-analysis. *British Journal of Psychiatry* 209, no. 6: 460–468. https://doi.org/10.1192/bjp.bp.115.175059

Korte, J., E.T. Bohlmeijer, P. Cappeliez, F. Smit, and G.F. Westerhof. 2012. Life review therapy for older adults with moderate depressive symptomatology: A pragmatic randomized controlled trial. *Psychological Medicine* 42, no. 6: 1163–1173.

Krause, K.R., D.B. Courtney, B.W.C. Chan, et al. 2021. Problem-solving training as an active ingredient of treatment for youth depression: A scoping review and exploratory meta-analysis. *BMC Psychiatry* 21, no. 1: 397. https://doi.org/10.1186/s12888-021-03260-9

Kudo, Y., A. Nakagawa, T. Wake, et al. 2017. Temperament, personality, and treatment outcome in major depression: A 6-month preliminary prospective study. *Neuropsychiatric Disease and Treatment* 13: 17–24.

Kupfer, D.J. 2005. The pharmacological management of depression. *Dialogues in Clinical Neuroscience* 7, no. 3: 191–205. https://doi.org/10.31887/DCNS.2005.7.3/dkupfer

Lam, R.W., A.J. Levitt, R.D. Levitan, et al. 2016. Efficacy of bright light treatment, fluoxetine, and the combination in patients with nonseasonal major depressive disorder: A randomized clinical trial. *JAMA Psychiatry* 73, no. 1: 56–63.

Lang, U.E., C. Beglinger, N. Schweinfurth, M. Walter, and S. Borgwardt. 2015. Nutritional aspects of depression. *Cellular Physiology and Biochemistry* 37, no. 3: 1029–1043. https://doi.org/10.1159/000430229

Lee, G., and H. Bae. 2017. Therapeutic effects of phytochemicals and medicinal herbs on depression. *BioMed Research International* 2017: 6596241. https://doi.org/10.1155/2017/6596241

Lee, J., M. Gierc, F. Vila-Rodriguez, E. Puterman, and G. Faulkner. 2021. Efficacy of exercise combined with standard treatment for depression compared to standard treatment alone: A systematic review and meta-analysis of randomized controlled trials. *Journal of Affective Disorders* 295: 1494–1511.

Li, M., X. Yao, L. Sun, et al. 2020. Effects of electroconvulsive therapy on depression and its potential mechanism. *Frontiers in Psychology* 11: 80. https://doi.org/10.3389/fpsyg.2020.00080

Lim, C.R., J. Barlas, and R.C.M. Ho. 2018. The effects of temperament on depression according to the schema model: A scoping review. *International Journal of Environmental Research and Public Health* 15, no. 6: 1231. https://doi.org/10.3390/ijerph15061231

Liu, L., C. Liu, Y. Wang, P. Wang, Y. Li, and B. Li. 2015. Herbal medicine for anxiety, depression and insomnia. *Current Neuropharmacology* 13, no. 4: 481–493. https://doi.org/10.2174/1570159x1304150831122734

Liu, Q., H. He, J. Yang, X. Feng, F. Zhao, and J. Lyu. 2020. Changes in the global burden of depression from 1990–2017: Findings from the global burden of disease study. *Journal of Psychiatric Research* 126: 134–140.

Liu, R.T., R.F.L. Walsh, and A.E. Sheehan. 2019. Prebiotics and probiotics for depression and anxiety: A systematic review and meta-analysis of controlled clinical trials. *Neuroscience and Biobehavioral Reviews* 102: 13–23. https://doi.org/10.1016/j.neubiorev.2019.03.023

Liu, Z., F. Yang, Y. Lou, W. Zhou, and F. Tong. 2021. The effectiveness of reminiscence therapy on alleviating depressive symptoms in older adults: A systematic review. *Frontiers in Psychology* 12: 709853. https://doi.org/10.3389/fpsyg.2021.709853

Lopizzo, N., L. Bocchio Chiavetto, N. Cattane, et al. 2015. Gene–environment interaction in major depression: Focus on experience-dependent biological systems. *Frontiers in Psychiatry* 6: 68. https://doi.org/10.3389/fpsyt.2015.00068

Maddock, A., and C. Blair. 2021. How do mindfulness-based programmes improve anxiety, depression and psychological distress? A systematic review. *Current Psychology.* https://doi.org/10.1007/s12144-021-02082-y

Madeeh Hashmi, A., M. Awais Aftab, N. Mazhar, M. Umair, and Z. Butt. 2013. The fiery landscape of depression: A review of the inflammatory hypothesis. *Pakistan Journal of Medical Sciences* 29, no. 3: 877–884. https://doi.org/10.12669/pjms.293.3357

Mahmud, S., M. Mohsin, M.N. Dewan, and A. Muyeed. 2022. The global prevalence of depression, anxiety, stress, and insomnia among general population during COVID-19 pandemic: A systematic review and meta-analysis. *Trends in Psychology.* https://doi.org/10.1007/s43076-021-00116-9

Majd Ara, E., S. Talepasand, and A.M. Rezaei. 2017. A structural model of depression based on interpersonal relationships: The mediating role of coping strategies and loneliness. *Noro Psikiyatri Arsivi* 54, no. 2: 125–130. https://doi.org/10.5152/npa.2017.12711

Makwana, N. 2019. Disaster and its impact on mental health: A narrative review. *Journal of Family Medicine and Primary Care* 8, no. 10: 3090–3095.

Maletic, V., M. Robinson, T. Oakes, S. Iyengar, S.G. Ball, and J. Russell. 2007. Neurobiology of depression: An integrated view of key findings. *International Journal of Clinical Practice* 61, no. 12: 2030–2040. https://doi.org/10.1111/j.1742-1241.2007.01602.x

Manalai, P., R.G. Hamilton, P. Langenberg, et al. 2012. Pollen-specific immunoglobulin E positivity is associated with worsening of depression scores in bipolar disorder patients during high pollen season. *Bipolar Disorders* 14, no. 1: 90–98.

March, J., S. Silva, S. Petrycki, et al. 2004. Fluoxetine, cognitive-behavioral therapy, and their combination for adolescents with depression: Treatment for adolescents with depression study (TADS) randomized controlled trial. *JAMA* 292, no. 7: 807–820.

Marchetti, I., T. Loeys, L.B. Alloy, and E.H.W. Koster. 2016. Unveiling the structure of cognitive vulnerability for depression: Specificity and overlap. *Plos One* 11, no. 2: e0168612. https://doi.org/10.137/journal.pone .0168612

Maruani, J., and P.A. Geoffroy. 2019. Bright light as a personalized precision treatment of mood disorders. *Frontiers in Psychiatry* 10: 85. https://doi.org/10.3389/fpsyt.2019.00085

Matta Mello Portugal, E., T. Cevada, R. Sobral Monteiro-Junior, et al. 2013. Neuroscience of exercise: From neurobiology mechanisms to mental health. *Neuropsychobiology* 68, no. 1: 1–14.

McEachan, R.R.C., S.L. Prady, G. Smith, et al. 2016. The association between green space and depressive symptoms in pregnant women: Moderating roles of socioeconomic status and physical activity. *Journal of Epidemiology and Community Health* 70: 253–259.

Melis, G., E. Gelormino, G. Marra, E. Ferracin, and G. Costa. 2015. The effects of the urban built environment on mental health: A cohort study in a large northern Italian city. *International Journal of Environmental Research and Public Health* 12, no. 11: 14898–14915.

Morganstein, J.C., and R.J. Ursano. 2020. Ecological disasters and mental health: Causes, consequences, and interventions. *Frontiers in Psychiatry* 11: 1. https://doi.org/10.3389/fpsyt.2020

Mullins, N., and C.M. Lewis. 2017. Genetics of depression: Progress at last. *Current Psychiatry Reports* 19: 43. https://doi.org/10.1007/s11920-017-0803-9

Murthy, R.S., and R. Lakshminarayana. 2006. Mental health consequences of war: A brief review of research findings. *World Psychiatry* 5, no. 1: 25–30.

Musisi, S., and E. Kinyanda. 2020. Long-term impact of war, civil war, and persecution in civilian populations-conflict and post-traumatic stress in African communities. *Frontiers in Psychiatry* 11: 20. https:// doi.org/10.3389/fpsyt.2020.00020

Mutz, J., V. Vipulananthan, B. Carter, R. Hurlemann, C.H.Y. Fu, and A.H. Young. 2019. Comparative efficacy and acceptability of non-surgical brain stimulation for the acute treatment of major depressive episodes in adults: Systematic review and network meta-analysis. *BMJ* 364: l1079. https://doi.org/10.1136/bmj .l1079

Nabeshima, T., and H.C. Kim. 2013. Involvement of genetic and environmental factors in the onset of depression. *Experimental Neurobiology* 22, no. 4: 235–243. https://doi.org/10.5607/en.2013.22.4.235

Nandam, L.S., M. Brazel, M. Zhou, and D.J. Jhaveri. 2020. Cortisol and major depressive disorder—Translating findings from humans to animal models and back. *Frontiers in Psychiatry* 10: 974. https:// doi.org/10.3389/fpsyt.2019.00974

National Collaborating Centre for Mental Health. 2010. *Depression in Adults with a Chronic Physical Health Problem: Treatment and Management*. British Psychological Society, Leicester. www.ncbi.nlm.nih.gov /books/NBK82926/

National Research Council (US) and Institute of Medicine (US). 2009. Depression, parenting practices, and the healthy development of children. In *Depression in Parents, Parenting, and Children: Opportunities to Improve Identification, Treatment, and Prevention*, edited by M.J. England, and L.J. Sim. National Academies Press, Washington, DC. https://www.ncbi.nlm.nih.gov/books/NBK215119/

Ndegwa, J. 2021. The efficacy of social skills training on depression among young adults in Nairobi county, Kenya. *Journal of Psychology* 3, no. 1: 18–28.

Necho, M., M. Tsehay, M. Birkie, G. Biset, and E. Tadesse. 2021. Prevalence of anxiety, depression, and psychological distress among the general population during the COVID-19 pandemic: A systematic review and meta-analysis. *International Journal of Social Psychiatry* 67, no. 7: 892–906.

Negele, A., J. Kaufhold, L. Kallenbach, and M. Leuzinger-Bohleber. 2015. Childhood trauma and its relation to chronic depression in adulthood. *Depression Research and Treatment* 2015: 650804. https://doi.org /10.1155/2015/650804

Nery, F.G., J.P. Hatch, M.A. Nicoletti, et al. 2009. Temperament and character traits in major depressive disorder: Influence of mood state and recurrence of episodes. *Depression and Anxiety* 26, no. 4: 382–388.

Nestler, E.J., M. Barrot, R.J. DiLeone, A.J. Eisch, S.J. Gold, and L.M. Montezgia. 2002. Neurobiology of depression. *Neuron* 34, no. 1: 13–25.

Nieuwsma, J.A., R.B. Trivedi, J. McDuffie, I. Kronish, D. Benjamin, and J.W. Williams. 2012. Brief psycho-therapy for depression: A systematic review and meta-analysis. *International Journal of Psychiatry in Medicine* 43, no. 2: 129–151. https://doi.org/10.2190/PM.43.2.c

Nogueira, B.S., R.F. Junior, I.M. Bensenor, P.A. Lotufo, and A.R. Brunoni. 2017. Temperament and character traits in major depressive disorder: A case control study. *Sao Paulo Medical Journal* 135, no. 5. https://doi.org/10.1590/1516-3180.2017.0063250517

Noonan, S., M. Zaveri, E. Macaninch, and K. Martyn. 2020. Food & mood: A review of supplementary pre-biotic and probiotic interventions in the treatment of anxiety and depression in adults. *BMJ Nutrition, Prevention & Health* 3, no. 2: 351–362. https://doi.org/10.1136/bmjnph-2019-000053

Ogbru, A. 2021. The comprehensive list of antidepressant medications. www.rxlist.com/the_comprehensive_list_of_antidepressants/drugs-condition.htm (accessed 28 July 2022).

Oldehinkel, A.J., R. Veenstra, J. Ormel, A.F. de Winter, and F.C. Verhulst. 2006. Temperament, parenting, and depressive symptoms in a population sample of preadolescents. *Journal of Child Psychology and Psychiatry, and Allied Disciplines* 47, no. 7: 684–695.

Owens, M., E. Watkins, M. Bot, et al. 2020. Nutrition and depression: Summary of findings from the EU-funded MooDFOOD depression prevention randomised controlled trial and a critical review of the literature. *Nutrition Bulletin* 45, no. 4: 403–414. https://doi.org/10.1111/nbu.12447

Ozbay, F., D.C. Johnson, E. Dimoulas, C.A. Morgan, D. Charney, and S. Southwick. 2007. Social support and resilience to stress: from neurobiology to clinical practice. *Psychiatry (Edgmont)* 4, no. 5: 35–40.

Padhy, S.K., S. Sarkar, M. Panigrahi, and S. Paul. 2015. Mental health effects of climate change. *Indian Journal of Occupational and Environmental Medicine* 19, no. 1: 3–7. https://doi.org/10.4103/0019-5278.156997

PAHO. n.d. Mental health. https://www.paho.org/en/topics/mental-health (accessed 25 July 2022).

Palaniyappan, L., L. Insole, and N. Ferrier. 2009. Combining antidepressants: A review of evidence. *Advances in Psychiatric Treatment* 15, no. 2: 90–99. https://doi.org/10.1192/apt.bp.107.004820

Palazidou, E. 2012. The neurobiology of depression. *British Medical Bulletin* 101, no. 1: 127–145.

Parekh, R. 2018. What is mental illness? www.psychiatry.org/patients-families/what-is-mental-illness (accessed 25 July 2022).

Parekh, R., and L. Givon. 2019. What is psychotherapy? https://www.psychiatry.org/patients-families/psycho-therapy (accessed 28 May 2022).

Parmentier, F.B.R., M. Garcia-Toro, J. Garcia-Campayo, A.M. Yanez, P. Andres, and M. Gili. 2019. Mindfulness and symptoms of depression and anxiety in the general population: The mediating roles of worry, rumination, reappraisal and suppression. *Frontiers in Psychology* 10: 506. https://doi.org/10.3389/fpsyg.2019.00506

Pascal Iloh, G.U., U.N. Orji, M.E. Chukwuonye, and C.V. Ifedigbo. 2018. The role of family bio-social variables in depression in a resource-constrained environment: A cross-sectional study of ambulatory adult patients in a primary care clinic in Eastern Nigerian. *Journal of Medical Sciences* 38: 29–37.

Patten, S.B., and E.J. Love. 1993. Can drugs cause depression? A review of the evidence. *Journal of Psychiatry and Neuroscience* 18, no. 3: 92–102.

Pelgrims, I., B. Devleesschauwer, M. Guyot, et al. 2021. Association between urban environment and mental health in Brussels, Belgium. *BMC Public Health* 21, no. 1: 635. https://doi.org/10.1186/s12889-021-10557-7

Preschl, B., A. Maercker, B. Wagner, et al. 2012. Life-review therapy with computer supplements for depression in the elderly: A randomized controlled trial. *Aging and Mental Health* 16, no. 8: 964–974.

Qin, D.D., J. Rizak, X.L. Feng, et al. 2016. Prolonged secretion of cortisol as a possible mechanism underlying stress and depressive behaviour. *Scientific Reports* 6: 30187. https://doi.org/10.1038/srep30187

Qin, P., B.L. Waltoft, P.B. Mortensen, and T.T. Postolache. 2013. Suicide risk in relation to air pollen counts: A study based on data from Danish registers. *BMJ Open* 3: e002462. https://doi.org/10.1136/bmjopen-2012-002462

Ragab, M.H., A.M. Alatik, M.M. el-Sha'abini, and A.S. Othman. 2000. Relationship between indoor environmental physical factors and depression aspects. *Journal of the Egyptian Public Health Association* 75, no. 3–4: 233–243.

Reavell, J., M. Hopkinson, D. Clarkesmith, and D.A. Lane. 2018. Effectiveness of cognitive behavioral therapy for depression and anxiety in patients with cardiovascular disease: A systematic review and meta-analysis. *Psychosomatic Medicine* 80, no. 8: 742–753.

Reeves, G.M., G.V. Nijjar, P. Langenberg, et al. 2012. Improvement in depression scores after 1 hour of light therapy treatment in patients with seasonal affective disorder. *Journal of Nervous and Mental Disease* 200, no. 1: 51–55.

Ribeiro, A., J.P. Ribeiro, and O. von Doellinger. 2018. Depression and psychodynamic psychotherapy. *Brazilian Journal of Psychiatry* 40, no. 1: 105–109. https://doi.org/10.1590/1516-4446-2016-2107

Rizvi, S., and A.M. Khan. 2019. Use of transcranial magnetic stimulation for depression. *Cureus* 11, no. 5: e4736. https://doi.org/10.7759/cureus.4736

Rogers, D., and R. Pies. 2008. General medical with depression drugs associated. *Psychiatry* 5, no. 12: 28–41.

Ruhe, H., N. Mason, and A. Schene. 2007. Mood is indirectly related to serotonin, norepinephrine and dopamine levels in humans: A meta-analysis of monoamine depletion studies. *Molecular Psychiatry* 12: 331–359. https://doi.org/10.1038/sj.mp.4001949

Salari, N., A. Hosseinian-Far, R. Jalali, et al. 2020. Prevalence of stress, anxiety, depression among the general population during the COVID-19 pandemic: A systematic review and meta-analysis. *Globalization and Health* 16: 57. https://doi.org/10.1186/s12992-020-00589-w

Salk, R.H., J.S. Hyde, and L.Y. Abramson. 2017. Gender differences in depression in representative national samples: Meta-analyses of diagnoses and symptoms. *Psychological Bulletin* 143, no. 8: 783–822. https://doi.org/10.1037/bul0000102

Samuelson, R. 2017. The impact of diet and nutrition on adolescent depression: A systematic review. Retrieved from Sophia, the St. Catherine University repository website: https://sophia.stkate.edu/msw_papers/786

Sanchez-Vidana, D.I., S.P. Ngai, W. He, J.K. Chow, B.W. Lau, and H.W. Tsang. 2017. The effectiveness of aromatherapy for depressive symptoms: A systematic review. *Evidence-Based Complementary and Alternative Medicine* 2017, no. 8: 1–21. https://doi.org/10.1155/2017/5869315

Sarris, J., A. Panossian, I. Schweitzer, C. Stough, and A. Scholey. 2011. Herbal medicine for depression, anxiety and insomnia: A review of psychopharmacology and clinical evidence. *European Neuropsychopharmacology* 21, no. 12: 841–860.

Schramm, E., S. Mack, N. Thiel, C. Jenkner, M. Elsaesser, and T. Fangmeier. 2020. Interpersonal psychotherapy vs. treatment as usual for major depression related to work stress: A pilot randomized controlled study. *Frontiers in Psychiatry* 11: 193. https://doi.org/10.3389/fpsyt.2020.00193

Segrin, C. 2000. Social skills deficit associated with depression. *Clinical Psychology Review* 20, no. 3: 379–403.

Setorki, M. 2020. Medicinal herbs with anti-depressant effects. *Journal of HerbMed Pharmacology* 9, no. 4: 309–317.

Shadrina, M., E.A. Bondarenko, and P.A. Slominsky. 2018. Genetics factors in major depression disease. *Frontiers in Psychiatry* 9: 334. https://doi.org/10.3389/fpsyt.2018.00334

Siegrist, J., and N. Wege. 2020. Adverse psychosocial work environments and depression- A narrative review of selected theoretical models. *Frontiers in Psychiatry* 11: 66. https://doi.org/10.3389/fpsyt.2020.00066

Singh, M.K., and I.H. Gotlib. 2014. The neuroscience of depression: Implications for assessment and intervention. *Behaviour Research and Therapy* 62: 60–73. https://doi.org/10.1016/j.brat.2014.08.008

Singh, N.A., K.M. Clements, and M.A. Singh. 2001. The efficacy of exercise as a long-term antidepressant in elderly subjects: A randomized, controlled trial. *Journals of Gerontology. Series A, Biological Sciences and Medical Sciences* 56, no. 8: M497–504. https://doi.org/10.1093/gerona/56.8.m497

Sohn, S.Y., P. Rees, B. Wildridge, N.J. Kalk, and B. Carter. 2019. Prevalence of problematic smartphone usage and associated mental health outcomes amongst children and young people: A systematic review, meta-analysis and GRADE of the evidence. *BMC Psychiatry* 19: 356. https://doi.org/10.1186/s12888-019-2350-x

Srivastava, K. 2009. Urbanization and mental health. *Industrial Psychiatry Journal* 18, no. 2: 75–76. https://doi.org/10.4103/0972-6748.64028

Steinert, C., T. Munder, S. Rabung, J. Hoyer, and F. Leichsenring. 2017. Psychodynamic therapy: As efficacious as other empirically supported treatments? A meta-analysis testing equivalence of outcomes. *American Journal of Psychiatry* 174, no. 10: 943–953.

Stoetzer, U., G. Ahlberg, G. Johansson, et al. 2009. Problematic interpersonal relationships at work and depression: A Swedish prospective cohort study. *Journal of Occupational Health* 51, no. 2: 144–151.

Sullivan, P.F., M.C. Neale, and K.S. Kendler. 2000. Genetic epidemiology of major depression: Review and meta-analysis. *American Journal of Psychiatry* 157, no. 10: 1552–1562.

Takahashi, T., F. Sugiyama, T. Kikai, et al. 2019. Changes in depression and anxiety through mindfulness group therapy in Japan: The role of mindfulness and self-compassion as possible mediators. *BioPsychoSocial Medicine* 13: 4. https://doi.org/10.1186/s13030-019-0145-4

Tao, L., R. Jiang, K. Zhang, et al. 2020. Light therapy in non-seasonal depression: An update meta-analysis. *Psychiatry Research* 291: 113247. https://doi.org/10.1016/j.psychres.2020.113247

Tasci, G., S. Baykara, M.G. Gurok, and M. Atmaca. 2019. Effect of exercise on therapeutic response in depression treatment. *Psychiatry and Clinical Psychopharmacology* 29, no. 2: 137–143.

Thair, H., A.L. Holloway, R. Newport, and A.D. Smith. 2017. Transcranial direct current stimulation (tDCS): A beginner's guide for design and implementation. *Frontiers in Neuroscience* 11: 641. https://doi.org/10.3389/fnins.2017.00641

Tolentino, J.C., and S.L. Schmidt. 2018. DSM-5 criteria and depression severity: Implications for clinical practice. *Frontiers in Psychiatry* 9: 450. https://doi.org/10.3389/fpsyt.2018.00450

Trofimova, I.N., and W. Sulis. 2016. A study of the coupling of FET temperament traits with major depression. *Frontiers in Psychology* 7: 1848. https://doi.org/10.3389/fpsyg.2016.01848

Tsehay, M., M. Necho, and W. Mekonnen. 2020. The role of adverse childhood experience on depression symptom, prevalence, and severity among school going adolescents. *Depression Research and Treatment* 2020: 5951792. https://doi.org/10.1155/2020/5951792

Uphoff, E., D. Ekers, L. Robertson, et al. 2020. Behavioural activation therapy for depression in adults. *Cochrane Database of Systematic Reviews* 7, no. 7: CD013305. https://doi.org/10.1002/14651858.CD013305.pub2

van den Bosch, M., and A. Meyer-Lindenberg. 2019. Environmental exposures and depression: Biological mechanisms and epidemiological evidence. *Annual Review of Public Health* 40: 239–259. https://doi.org/10.1146/annurev-publhealth-040218-044106

van der Velden, A.M., W. Kuyken, U. Wattar, et al. 2015. A systematic review of mechanisms of change in mindfulness-based cognitive therapy in the treatment of recurrent major depressive disorder. *Clinical Psychology Review* 37: 26–39. https://doi.org/10.1016/j.cpr.2015.02.001

Van Dooren, F.E.P., F.R.J. Verhey, F. Pouwer, et al. 2016. Association of Type D personality with increased vulnerability to depression: Is there a role for inflammation or endothelial dysfunction?- The Maastricht study. *Journal of Affective Disorders* 189: 118–125.

Verduijn, J., Y. Milaneschi, R.A. Schoevers, A.M. van Hemert, A.T. Beekman, and B.W. Penninx. 2015. Pathophysiology of major depressive disorder: Mechanisms involved in etiology are not associated with clinical progression. *Translational Psychiatry* 5, no. 9: e649. https://doi.org/10.1038/tp.2015.137

Wagner, K.D. 2016. Effects of childhood trauma on depression and suicidality in adulthood. *Psychiatric Times* 33, no. 11. www.psychiatrictimes.com/view/effects-childhood-trauma-depression-and-suicidality-adulthood (accessed 20 February 2022).

Wallace, C.J.K., and R.V. Milev. 2021. The efficacy, safety, and tolerability of probiotics on depression: Clinical results from an open-label pilot study. *Frontiers in Psychiatry* 12: 618279. https://doi.org/10.3389/fpsyt.2021.618279

Wemrell, M., A. Olsson, and K. Landgren. 2020. The use of complementary and alternative medicine (CAM) in psychiatric units in Sweden. *Issues in Mental Health Nursing* 41, no. 10: 946–957.

Whiteford, H.A., L. Degenhardt, J. Rehm, et al. 2013. Global burden of disease attributable to mental and substance use disorders: Findings from the Global Burden of Disease Study 2010. *Lancet* 382, no. 9904: 1575–1586.

WHO. 2021a. Depression. https://www.who.int/news-room/fact-sheets/detail/depression (accessed 20 February 2022).

WHO. 2021b. *WHO Model List of Essential Medicines-22nd List, 2021.* WHO, Geneva (accessed 20 May 2022).

Woo, J.M., and T.T. Postolache. 2008. The impact of work environment on mood disorders and suicide: Evidence and implications. *International Journal on Disability and Human Development* 7, no. 2: 185–200. https://doi.org/10.1515/ijdhd.2008.7.2.185

Wu, Y., J. Mo, L. Sui, et al. 2021. Deep brain stimulation in treatment-resistant depression: A systematic review and meta-analysis on efficacy and safety. *Frontiers in Neuroscience* 15: 655412. https://doi.org/10.3389/fnins.2021.655412

Xie, Y., Z. Wu, L. Sun, et al. 2021. The effects and mechanisms of exercise on the treatment of depression. *Frontiers in Psychiatry* 12: 705559. https://doi.org/10.3389/fpsyt.2021.705559

Yan, C., H. Liao, Y. Ma, and J. Wang. 2022. Association amongst social support inside or outside the family and depression symptoms: Longitudinal study of urban–rural differences in China. *Quality of Life Research* 31, no. 6: 1677–1687. https://doi.org/10.1007/s11136-022-03086-2

Yeung, K.S., M. Hernandez, J.J. Mao, I. Haviland, and J. Gubili. 2018. Herbal medicine for depression and anxiety: A systematic review with assessment of potential psycho-oncologic relevance. *Phytotherapy Research* 32, no. 5: 865–891. https://doi.org/10.1002/ptr.6033

Yu, M., K.A. Linn, R.T. Shinohara, et al. 2019. Childhood trauma history is linked to abnormal brain connectivity in major depression. *Proceedings of the National Academy of Sciences of the United States of America* 116, no. 17: 8582–8590.

Yu, Y., X. Yang, Y. Yang, et al. 2015. The role of family environment in depressive symptoms among university students: A large sample survey in China. *PLoS One* 10, no. 12: e0143612. https://doi.org/10.1371/journal.pone.0143612

Zappitelli, M.C., I.A. Bordin, J.P. Hatch, et al. 2013. Temperament and character traits in children and adolescents with major depressive disorder: A case-control study. *Comprehensive Psychiatry* 54, no. 4: 346–353.

Zhang, A., S. Park, J.E. Sullivan, and S. Jing. 2018. The effectiveness of problem-solving therapy for primary care patients' depressive and/or anxiety disorders: A systematic review and meta-analysis. *Journal of the American Board of Family Medicine* 31, no. 1: 139–150.

Zhichao, H., L.W. Ching, L. Huijuan, et al. 2021. A network meta-analysis on the effectiveness and safety of acupuncture in treating patients with major depressive disorder. *Scientific Reports* 11, no. 1: 10384. https://doi.org/10.1038/s41598-021-88263-y

10 Neurocognitive Disorder (Dementia)

10.1 INTRODUCTION

Dementia is a syndrome leading to deterioration in cognitive functioning. It is usually chronic or progressive in nature and affects memory, thinking, orientation, calculation, learning capacity, judgement, attention, language, and comprehension (WHO 2021; Arvanitakis et al. 2019). According to the International Classification of Diseases (ICD)-10, dementia (F00–F03) is a syndrome due to diseases of the brain in which there is a disturbance of multiple higher cortical functions. The impairment in cognitive functioning is accompanied and occasionally preceded by deterioration in emotional control or social behaviour or motivation (WHO 2016).

According to Diagnostic and Statistical Manual of Mental Disorders (DSM)-IV, 'delirium, dementia, and amnestic and other cognitive disorders' are grouped under clinical disorders in Axis 1, wherein dementia is described as a condition characterised by the development of multiple cognitive deficits due to the direct physiological effects of a generalised medical condition, the persisting effects of a substance, or multiple aetiologies (Sachdev et al. 2014) but in 2013, DSM-5 replaced 'delirium, dementia, and amnestic and other cognitive disorders' with neurocognitive disorders. In DSM-5, the major change is the replacement of the term dementia with neurocognitive disorders. The major neurocognitive disorder is used as a synonym for dementia (Hugo and Ganguli 2014; Siberski 2012). In DSM-5, the neurocognitive disorder cluster is divided into three syndromes, namely, delirium, mild neurocognitive disorder, and major neurocognitive disorder, as shown in Figure 10.1 (Sachdev et al. 2014).

10.2 DIAGNOSTIC CRITERIA FOR NEUROCOGNITIVE DISORDERS

As we know, neurocognitive disorders are classified into three categories, namely delirium, mild neurocognitive disorders, and major neurocognitive disorders; therefore it is essential to understand the criteria for diagnosing each category of neurocognitive disorder.

Delirium is diagnosed as a disturbance in attention and awareness developed over a short period of time. Disturbance in cognition is also present. This disturbance is not because of other pre-existing, established, or evolving neurocognitive disorders and is the direct physiological consequence of another medical condition, substance intoxication or withdrawal, or exposure to toxins or is due to multiple aetiologies (Sachdev et al. 2014).

The diagnostic criteria for mild neurocognitive disorder are a modest decline in one or more cognitive domains with no interference with everyday activities, but the activities require more time and effort. In major neurocognitive disorder (dementia), a substantial decline in one or more cognitive domains occurs and the impairment is sufficient to interfere with independence in everyday activities. The other criteria for major and mild cognitive impairment are that the cognitive deficits do not occur in the context of delirium; the cognitive deficits are not primarily attributed to other mental disorders, and further, specify one or more etiological subtypes (Hugo and Ganguli 2014; Siberski 2012).

There are six cognitive domains, namely, learning and memory, social cognition, complex attention, executive function, perceptual-motor function, and language. Any decline from the previous

DOI: 10.1201/9781003354024-10

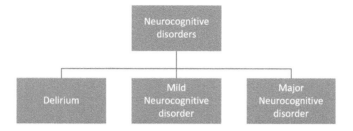

FIGURE 10.1 Categorisation of neurocognitive disorders.

level of functioning in any of these key cognitive domains forms the criteria for diagnosing neuro-cognitive disorders (Sachdev et al. 2014).

10.3 PREVALENCE

Dementia is the seventh leading cause of death and one of the major causes of disability and dependency. Currently, more than 55 million people worldwide live with dementia and there are nearly 10 million new cases every year. This number is expected to increase to 78 million in 2030 and 139 million in 2050. The gravity of the problem may be understood based on the basis of an estimate stating that there are 10 million new cases of dementia each year (WHO 2021). It has also been estimated that the increase in cases of dementia will be from 57.4 million in 2019 to 152.8 million in 2050, with the highest projected increase in North Africa, the Middle East, and eastern sub-Saharan Africa regions (Alzheimer's Association 2021), indicating a greater increase in countries with a low socioeconomic index. It is expected that a larger increase in cases of dementia will be seen in middle- and low-income countries in the coming decades probably because the risk factors such as hypertension, diabetes, and cardiovascular conditions will increase in these countries. The increase will also be due to an increase in the elderly population and population growth. The prevalence of dementia was observed to be higher in females compared to males in 2019, and the same trend is expected in 2050 (GBD 2019; Dementia Forecasting Collaborators 2022).

10.4 AETIOLOGY/RISK FACTORS

Dementia may be caused by brain disease or brain injury or medical conditions of other body parts. Getting old may be a risk factor for some, though dementia is not necessarily part of ageing (Arvanitakis et al. 2019). The aetiological/risk factors for dementia may be categorised as non-modifiable and modifiable factors as shown in Figure 10.2.

10.4.1 Non-modifiable Factors

10.4.1.1 Genetic Influence

Dementias are proteinopathies and some dementia cases may be familial, i.e., caused by dominant-acting disease gene mutation (Paulson and Igo 2011). It has been seen that the risk of developing dementia increased monotonically across the genetic risk, independent of lifestyle factors. The higher the genetic risk, the greater is the percentage of individuals developing dementia (Lourida et al. 2019).

10.4.1.2 Age

Dementia usually is a disorder of old age, i.e., the risk is higher after 65 years. However, 2–10% of all cases may initiate before 65 years. It has been reported that after 65 years of age, the prevalence of dementia doubles with an increase in 5 years of age (World Alzheimer Report 2014). The

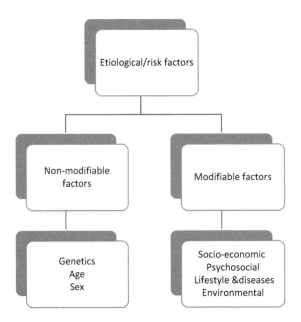

FIGURE 10.2 Aetiological/risk factors for dementia.

prevalence and incidence rate of dementia increase exponentially with advancing age (van der Flier and Scheltens 2005).

10.4.1.3 Sex

Women are at a higher risk of dementia than men. It has been reported that worldwide women with dementia outnumber men by two times. One of the probable reasons is that women live longer than men and since dementia is a disorder of age, the proportion of women is higher than men. Another reason influencing the sex difference in dementia is the fluctuation in hormones at different life stages in women, especially menopause. The hormone oestrogen is postulated to modulate the effect of the (apolipoprotein E (APOE) gene E4 allele. The E4 allele of the APOE is the risk factor for developing dementia (Rocca et al. 2014). Moreover, females have fewer modifiable risk factors associated with memory decline than males, increasing the prevalence of dementia in women. Demographic and lifestyle domains affect men more than women, whereas genetic and medical domains affect women more than men (Anstey et al. 2021). It has been seen that women show faster decline in cognition and executive functions, bringing them at a higher risk of developing dementia (Levine et al. 2021).

10.4.2 MODIFIABLE RISK FACTORS OVER THE LIFESPAN

10.4.2.1 Socioeconomic Status

Socioeconomic indicators like wealth, education, and employment status have been found to be correlated with dementia. Low educational attainment has been observed to be directly associated with dementia and dementia-related death, especially in women (Russ et al. 2013). It has been seen that dementia risk increased by 3–4 fold in individuals who had six years or less of education compared to those who had ten years or more of education (Nakahori et al. 2018). Higher education levels lead to higher cognitive reserves, delay cognitive decline, and enable individuals to be more resilient to cognitive decline. Higher cognitive reserves in individuals with a higher level of education meant that they have better reserves for language performance, visual-spatial ability, memory, and executive ability, thereby facilitating them to cope with cognitive declines (Chen et al. 2019). Thus, a low education level becomes a risk factor for dementia.

Like education, wealth also has been found to be related to dementia (Cadar et al. 2018; Deckers et al. 2019). Financial strain has been found to be associated with dementia (Samuel et al. 2020).

Employment status also modulates the risk factor for dementia. Unemployed individuals are at a higher risk of developing dementia compared to their employed counterparts. Furthermore, the type of employment also impacts the rate of developing dementia. More work control reduces the risk of dementia in men, as it could moderate the effect of the APOE E4 allele, a major genetic risk factor for dementia (Hasselgren et al. 2018). Involvement in an occupation requiring a higher complexity of work lowers the risk of dementia, probably because of increased mental stimulation due to intellectually demanding tasks (Kroger et al. 2008) and on the other hand, individuals working night shifts for a long period may be at risk of developing dementia, probably as it may disturb the normal circadian rhythm and lead to sleep disturbances leading to cognitive damaging effects (Leso et al. 2021). Employment status has also been seen to have a profound effect on neuroplasticity and cognitive reserve. Active engagement in work helps to learn new skills, increases social engagement, establishes a routine, gives a purpose and source of income. These five mechanisms support cognitive reserves (Vance et al. 2016). Paid work has a significant positive effect on cognitive abilities (Madhvan et al. 2022).

Lack of education and unemployment as major dimensions of poverty have been found to be associated with a higher prevalence of dementia (Trani et al. 2022), more so in individuals aged 60–74 (Ong et al. 2021) and these sociodemographic variables, along with others like rural residence and unmarried marital status, increased the rate of prevalence of dementia (Hamid et al. 2010). Higher education and occupations with high complexity level have a protective effect against dementia and have been found to have 7.1 or 4.6 times better global cognitive function in individuals who attained higher education or worked in occupations with a high complexity level respectively, compared to their counterparts (Darwish et al. 2018). Thus, it can be said that education level, economic status, employment status, marital status, and place of residence are some of the factors that may pose a risk for dementia.

10.4.2.2 Psychosocial Factors

Certain psychological conditions like depression, anxiety, psychological distress, and sleep disorders have been found to be associated with dementia. Depression can be a causal risk factor for dementia as depression is related to the release of pro-inflammatory cytokines, increased production of glucocorticoids, deposition of amyloid, and formation of neurofibrillary resulting in hippocampal injury. However, depression can also be a consequence of dementia (World Alzheimer Report 2014; Russ et al. 2011). It has been seen that individuals with a mental disorder have an increased risk of developing dementia. Subjects diagnosed with psychosis, substance abuse, mood disorder, neurosis, and all other mental disorders are more likely to develop dementia than those without mental disorders. Individuals who were diagnosed with a mental disorder developed dementia 5.6 years earlier than those subjects who were not diagnosed with a mental disorder (Richmond-Rakerd et al. 2022). Depression, higher anxiety, loneliness, social isolation, and lack of engagement in mental activities are associated with the risk of developing dementia (PSSRU 2017). An association between the presence of psychiatric disorder and dementia was noted by Tori et al. (2020).

10.4.2.3 Lifestyle Factors and Diseases

An unhealthy lifestyle increases the risk of dementia (Deckers et al. 2019). Alongside an unhealthy lifestyle, certain medical conditions increase the susceptibility to dementia. Dementia is associated with alcohol consumption, smoking, diabetes, Parkinson's disease, stroke, and angina pectoris/cardiovascular diseases (Nakahori et al. 2018). A favourable lifestyle is crucial for preventing dementia and it has been seen that a healthy lifestyle is associated with a lower risk of dementia, irrespective of genetic risk. On the other hand, a higher percentage of individuals following an unhealthy lifestyle develop dementia compared to those with a healthy lifestyle (Lourida et al. 2019).

Physical inactivity, current smoking, and heavy and chronic drinking are associated with all dementia risks (PSSRU 2017). Subjects who lead a sedentary life are at a higher risk of developing dementia compared to those who are physically active. Heavy alcohol consumers and smokers are at a higher risk of developing dementia; however, conclusive inference may be drawn after analysing the duration, frequency, and type of alcohol consumption and smoking habits along with other associated risks (Di Marco et al. 2014).

Nutrients are important for the normal functioning of the body. Certain nutrients play an important role in regulating the activities in the brain and thus, their deficiency may lead to dementia. Vitamin D deficiency is observed as a risk factor for dementia, probably because of its role in clearing amyloid β (Aβ) protein and maintaining calcium homeostasis, thereby having a protective effect against neuronal ageing and neurodegeneration (Chai et al. 2019). Deficiency of vitamin B12 and folic acid may also act as a risk factor for dementia, possibly because their deficiency leads to an increase in homocysteine level which may contribute to the accumulation of Aβ and tau protein and consequently neuronal death (Pathak and Mattos 2021). In a systematic review, it has been found that higher values of total dietary fat, trans unsaturated fats, and total calorie intake are associated with an increased risk of dementia (Di Marco et al. 2014). The intrauterine environment and the first two years of life are crucial in the development of the brain, therefore a poor intrauterine environment, placental function, and maternal malnutrition may have an impact on cognitive functioning. The brain and skull grow rapidly until 6 years of age and attain their maximum growth by 15 years of age. Proper nutrition during childhood and adolescence helps in maintaining good brain and cognitive reserves, thereby acting as a buffer for expression of symptoms of dementia (World Alzheimer Report 2014).

Diseased states present a risk of dementia. During adulthood, cardiovascular risks like high blood pressure, cholesterol, diabetes, overweight, and obesity may increase the risk of the development of dementia (World Alzheimer Report 2014). Stroke and hypertension are related to memory decline in men and women, respectively (Anstey et al. 2021). Subjects reporting poor health status and stroke show a higher rate of prevalence of dementia compared to their counterparts who reported good health status and no incidence of stroke (Hamid et al. 2010). In a study, it was concluded that modifiable lifestyle risk factors, such as midlife hypertension, midlife obesity, diabetes mellitus, physical inactivity, smoking, and low educational attainment, accounted for more than 50% of dementia cases, with the maximum proportion attributed to midlife hypertension (Ashby-Mitchell et al. 2018). Diseased states such as midlife obesity, midlife hypertension, and diabetes increase the risk of developing dementia (PSSRU 2017). Diabetes mellitus doubled the risk of dementia (Ott et al. 1999). In patients with diabetes mellitus, the risk of developing all types of dementia has been observed to increase by 73% compared to non-diabetic subjects (Gudala et al. 2013).

10.4.2.4 Environmental Factors

Exposure to environmental risk factors like air pollution, heavy metals, other metals, trace elements, pesticides, and electromagnetic fields, leads to cognitive impairment and may increase the risk of dementia. However, more studies are required for a definitive conclusion (Killin et al. 2016; Zhao et al. 2021).

10.5 SIGNS AND SYMPTOMS

Neuropsychiatric symptoms, also known as behavioural and psychological symptoms of dementia (BPSD), are present in virtually all patients suffering from dementia. These symptoms affect up to 90% of subjects during the course of the illness. The symptoms include disturbances in emotional experience; delusion and abnormal thought content; perceptual disturbances; motor function disturbances; circadian rhythm disturbances; and disturbances in appetite and eating behaviour. The most common symptoms are apathy, depression, irritability, agitation, and anxiety. Euphoria, hallucinations, and disinhibition are rarely present. Clinically significant symptoms include depression, apathy, and anxiety. It has been observed that 50% of patients have at least four neuropsychiatric symptoms simultaneously (Cerejeira et al. 2012).

Short-term memory problems like repeatedly forgetting things that have happened in the recent past, difficulty with daily tasks, withdrawal from hobbies and social events, anxiety, and depression (Arvanitakis et al. 2019) are also seen. WHO (2021) has classified the signs and symptoms of dementia depending on the stage of severity of the illness. Forgetfulness, losing track of time, and becoming lost in familiar places are symptoms of the early stage. Forgetting recent events and names of people, confusion while at home, difficulty in communication, needing help with personal care, and behavioural changes can be seen during the middle stage. The last stage of dementia is characterised by complete dependency and inactivity, difficulty in recognising relatives and friends, unawareness of the time and place, increased need for assistance in self-care, difficulty in walking, and behavioural changes.

10.6 SUBCLASSIFICATION BY AETIOLOGY

Dementia is not a normal part of ageing and occurs due to a variety of brain illnesses. The aetiological subtypes are Alzheimer's disease, vascular neurocognitive disorder, frontotemporal neurocognitive disorder, Lewy body dementia, Parkinson's disease, traumatic brain injury, HIV infection, Huntington's disease, prion disease, and neurocognitive disorder due to any other medical condition. The categorisation depends on whether the cognitive impairment is caused by a disorder or if the cognitive decline led to some disorder subtypes. The decline in cognitive functioning can be due to some disorder like Parkinson's disease, HIV infection, Huntington's disease, traumatic brain injury, or, in some cases the cognitive and behavioural symptoms are evident first, followed by a longitudinal course that reveals aetiologies like Alzheimer's disease, cerebrovascular disease, frontotemporal lobar degeneration, and Lewy body disease (Sachdev et al. 2014).

The clinical syndrome of dementia can be due to varied pathophysiological processes. The most common are Alzheimer's disease accounting for 50–75% of cases of dementia, vascular dementia accounting for 20–30%, dementia with Lewy bodies accounting for <5%, and frontotemporal dementia accounting for 5–10% of dementia cases (World Alzheimer Report 2014). The most common forms of dementia are discussed in this chapter.

10.6.1 ALZHEIMER'S DISEASE (AD)

Alzheimer's disease is a brain disorder that slowly destroys memory and thinking skills and, eventually, the ability to carry out the simplest tasks. This damage initially takes place in parts of the brain involved in memory, including the entorhinal cortex and hippocampus. It later affects areas in the cerebral cortex, such as those responsible for language, reasoning, and social behaviour. Eventually, many other areas of the brain are damaged (National Institute of Aging 2021).

Alzheimer's disease is the most common cause of dementia. It is a degenerative disease caused by aberrant processing and polymerisation of soluble neuronal proteins. The extracellular aggregates of Aβ plaques and intracellular aggregation of neurofibrillary tangles are the pathological causes of AD. The Aβ plaques develop initially in the basal, temporal, and orbitofrontal neocortex regions of the brain and later stages spread throughout the neocortex, hippocampus, amygdala, diencephalon, and basal ganglia. Amyloid precursor protein gets cleaved by β-secretases and γ-secretases to produce insoluble Aβ fibrils, which further polymerise to form plaques. The polymerisation leads to the activation of kinase, which in turn leads to hyperphosphorylation of the microtubule-associated tau (τ) protein, and further polymerisation of it leads to the formation of neurofibrillary tangles. Aβ and neurofibrillary tangles interfere with neurone-to-neurone communication at synapses, resulting in synaptic damage, increased reactive oxidative stress, and consequently neuronal dysfunction leading to AD. Aβ plaques induce mitochondrial damage, unstable homeostasis, and synaptic dysfunction. Neurofibrillary tangles reduce the number of synapses, produce neurotoxicity, and cause cell dysfunction (Tiwari et al. 2019; Fan et al. 2020).

Neurones in the part of the brain involved in cognitive functions, such as memory, thinking, and learning get damaged resulting in the manifestation of symptoms. As the degeneration progresses, the neurones in other parts of the brain also get damaged, hampering the conduct of basic bodily functions. AD accounts for 60–80% of dementia cases. It is a slow progressive disease that begins many years before symptoms start manifesting. The progression of AD occurs in three phases: Preclinical AD, Mild Cognitive Impairment due to AD, and Dementia due to AD (Alzheimer Association Report 2020).

a. *Preclinical AD.* In this phase, measurable changes in the brain like abnormal levels of Aβ and decreased metabolism of glucose (fuel for brain) is observed. However, symptoms like memory loss do not develop by this time. The person is able to perform functions normally.
b. *Mild Cognitive Impairment due to AD.* In this phase, besides the measurable changes in the brain, subtle problems with memory and thinking occur, which are not quite noticeable, as the individual is able to carry out his daily functions.
c. *Dementia due to AD.* In this phase, noticeable changes in memory, thinking, and behaviour occur and the individual is not able to carry out daily activities with an increase in severity (Alzheimer Association Report 2020).

Depending on the age of onset of the disease, AD is classified into two categories: early-onset AD and late-onset AD. Cases of AD before 65 years of age are considered as early-onset AD. Early-onset AD is also referred to as familial AD because they are often caused by mutations in specific genes that cause autosomal dominant dementia in the family. The three genes implicated in AD are amyloid precursor protein gene, presenilins 1, and presenilins 2. Mutation in these three genes, especially presenilins 1, leads to familial AD; however, they comprise only 2% of all AD (Paulson and Igo 2011).

Late-onset AD is not a genetic disorder but genes may play an important risk factor. The APOE E4 allele is associated with an increased risk of late-onset AD. A few more genes, namely CLU, CR1, PICAM, BIN1, ABCA7, MS4A, CD33, and CD2AP have been identified as potent genetic risk factors contributing to the pathogenesis of late-onset AD, probably by influencing the immune system, neuroinflammation, synaptic dysfunction, cholesterol homeostasis dysregulation, and endocytosis (Paulson and Igo 2011).

The risk factors for AD are old age, female gender, presence of apolipoprotein gene E4 allele (APOE4), family history of AD, low education and occupational attainment, racial disparities, traumatic brain injury, and cardiovascular comorbidities (Apostolova 2016; Qiu et al. 2009).

The early clinical symptoms include difficulty in remembering recent conversations, names or places, apathy, and depression. The later symptoms consist of impaired communication, disorientation, confusion, poor judgement, behavioural changes, and ultimately difficulty in walking, swallowing, and speaking leading to complete dependency. The symptoms emerge years after the disease actually begins, as AD is slow and progressive in nature (Alzheimer's Association Report 2020). Memory impairment, aphasia, a decline in executive functions, and visual-spatial skills are the features of cognitive decline in AD. Additionally, neuropsychiatric symptoms such as apathy, anxiety, irritability, agitation, sundowning, psychosis, and diminished insight can be observed. Depression, sleep disorders, and a change in appetite are also evident.

According to the National Institute of Aging and the Alzheimer's Association, the diagnostic criteria for probable AD dementia are: the diagnosis for dementia should be established; onset is insidious with gradual progression; initial symptoms, amnestic and non-amnestic; no other neurologic, psychiatric, or general medical disorder of severity interfering with cognition is present; and the diagnostic certainty is established with positive biomarkers (Apostolova 2016).

10.6.2 Vascular Dementia (VD)

It is the second most common aetiological subtype of dementia after AD. VD, as the name suggests, occurs as a result of disruption in the flow of blood (oxygen) to the brain, leading to cognitive impairments and finally dementia.

Vascular dementia is a disease with cognitive impairment resulting from cerebrovascular disease, ischemic or brain haemorrhage. However, sometimes it is not necessary that an individual will necessarily fulfil the criteria of dementia, and only cognitive impairment is associated with cerebrovascular disease. Therefore, because of this condition, the older term vascular dementia is replaced with vascular cognitive impairment (VCI), in order to cover the entire spectrum of cognitive impairment caused by vascular factors. VD can be seen as the most severe form of VCI (Iemolo et al. 2009; Iadecola et al. 2019). In simpler terms, we can say that VCI includes cognitive impairments due to cerebrovascular disease, ranging from mild cognitive impairment to vascular dementia.

According to the Vascular Impairment of Cognition Classification Consensus Study guidelines, the first requirement for Major VCI (VD) is a significant decrease in one of the cognitive domains, causing severe disruption in daily activities. The second is imaging evidence of cerebrovascular disease. Neuroimaging may assess brain atrophy, white matter hyperintensities, infarction, and haemorrhage. Memory decline is not an essential criterion for VD, as in case with AD. Impairment in executive function, processing speed, and delayed recall of word lists and visual content is seen in VCI (Iadecola et al. 2019).

VD is classified into the following major subtypes: 1) post-stroke dementia, 2) subcortical ischemic vascular dementia, 3) multi-infarct dementia, and 4) mixed dementia (Iadecola et al. 2019). Cerebral autosomal dominant arteriopathy with subcortical infarcts and leukoencephalopathy (CADASIL) is yet another subtype of VD. The classification is based on the root cause of VD. Stroke-related dementia occurs when the blood supply to a part of the brain is suddenly interrupted because of narrowing of blood vessels, blockage by a clot, or rupture of blood vessels. Subcortical dementia is the most common type of VD and is caused due to reduced blood flow in the small blood vessels lying deep in the brain. Multi-infarct dementia is caused by a series of mini strokes also known as a transient ischemic attack. These multiple clots lead to the death of small area of the brain known as an infarct. Mixed dementia is a combination of VD with AD, which means the subject has symptoms of both VD and AD (Nazarko 2019). The chances of developing dementia increase two-fold after having stroke. Multi-infarct dementia occurs when a series of strokes leads to the loss of brain function. In subcortical VD, lacunar infarct and ischemic white matter lesions with demyelination and axon loss are observed, occurring primarily due to an injury to the penetrating arteries vessel walls (Akhter et al. 2021). CADASIL, a genetically determined arteriopathy, is a rare cause of VD. CADASIL, an early onset-disease, is caused by a mutation of the *Notch 3* gene on chromosome 19. The evident signs of CADASIL are attacks of migraine with aura, recurrent subcortical stroke, mood disturbances, and a progressive cognitive decline leading to dementia. Executive and organising cognitive functioning are affected followed by memory decline in CADASIL. The smooth muscle cells in small arteries get degenerated and the vessel walls become fibrotic resulting in circulatory disturbances and lacunar infarcts mainly in cerebral white matter and deep grey matter (Kalimo et al. 2008; Choi 2010).

On the basis of the pattern of vascular brain lesions leading to dementia, VD is distinguished into three forms: 1) multi-infarct dementia, 2) strategic infarct dementia, and 3) subcortical vascular encephalopathy. In multi-infarct dementia, multiple microinfarcts, lacunar infarcts, and small large infarcts are distributed all over the grey matter, and in strategic infarct dementia, infarct is seen in the thalamus region. In subcortical vascular encephalopathy, confluent white matter lesions are seen in the central and peripheral white matter (Jellinger 2013). Major stroke, lacunar stroke, microinfarcts, diffuse subcortical small vessel disease, and cerebral amyloid angiopathy are seen as major pathologies contributing to vascular cognitive impairment or VD. Small vessel disease, an important contributor to VD, affects the frontal lobe, thereby influencing the executive functioning leading to cognitive dysfunction (Series and Esiri 2012).

Vessel disorders like atherosclerosis, small vessel disease, and cerebral amyloid angiopathy cause cerebrovascular lesions leading to impairment in brain functioning. These disorders occur more frequently with advancing age and their severity leads to VD (Jellinger 2013).

To sum up the pathogenesis, it can be said that certain modifiable and non-modifiable risk factors lead to small and large vessel injury resulting in macroinfarct, lacunar infarct, microinfarct, and white matter changes. These changes contribute to brain atrophy, thereby VD (Bir et al. 2021).

Environmental factors like air quality, toxic metals, trace elements, work hazards, and weather patterns are also linked with VD (Akhter et al. 2021). Certain risk factors like advancing age, genetics, stroke, AD, hypertension, orthostatic hypotension, cardiac disease, diabetes, obesity, smoking, major surgery, elevated homocysteine levels, and hyperlipidaemia, APOE E4, and low education levels increase the susceptibility for vascular problems like atherosclerosis, microvascular diseases, reduced cerebral blood flow, endothelial disorders, oxidative stress, and blood brain barrier dysfunction leading to cerebral infarcts, lacunar infarcts, white matter lesions, multiple microinfarcts which eventually result in cognitive impairment and dementia (Series and Esiri 2012; Jellinger 2013). Genetic factors act as causal factors for VD in two ways. Firstly, the genes responsible for increasing the susceptibility of cerebrovascular diseases (hypertension, stroke) and secondly, the genes determining the tissue responses to cerebrovascular diseases, i.e., genes responsible for recovering from ischemic injury, increase the susceptibility of developing VD. Also, genetic factors may indirectly involve with the conventional risk factors such as hypertension, diabetes, and homocysteine concentration, thereby contributing to VD (Iemolo et al. 2009).

VD, the second most common form of dementia after AD, is more prevalent in developing nations. VD cases range from 15 to 20% in North America and Europe to 30% in Asia and developing countries. The risk of dementia increases by 7% in people with a first-ever stroke to 40% in people having recurrent stroke, indicating a rise in risk with a rise in severity and recurrence of stroke. The incidence of VD increases steeply with the increase in age (Wolters and Ikram 2019). Considering the autopsy-verified cases, the frequency of VD is estimated to be 10–15% globally (Kalaria 2016).

The symptoms of VD depend on the area of the brain affected. The common symptoms are disorientation, difficulty in thinking, inability to create new memories, agitation, loss of executive functions, memory impairment, and behavioural symptoms (Kalaria 2016; Akhter et al. 2021), deficit in attention and processing of information, difficulties in complex activities, and disorganised thoughts and behaviour. Depression and apathy are also common. Hallucinations and delusions are less common (Bir et al. 2021). Visuospatial function, semantic memory, affective and vegetative functioning were affected along with depression and poor sleep quality in subjects with multi-infarct dementia compared to healthy counterparts (Al-Adawi et al. 2014).

10.6.3 LEWY BODY DEMENTIA (LBD)

Lewy body dementia is the third common cause of dementia after AD and VD. It is a progressive degenerative disorder affecting the older population. LBD is a proteinopathy caused by abnormal accumulation of a protein called α-synuclein in the neuronal and non-neuronal cells in the brain. The clump of protein called Lewy bodies and Lewy neurites cause damage to areas of the brain associated with mental capabilities, behaviour, movement, and sleep by promoting oxidative damage and an inflammatory response leading to cell death. The α-synuclein alterations affect the monoaminergic systems, dysregulating acetylcholine and dopamine levels. Acetylcholine is important for memory and learning, and dopamine is essential for behaviour, cognition, movement, sleep, and mood regulation. A deficiency of acetylcholine in the temporal and parietal cortex results in hallucination, and upregulation of muscarinic M1 receptors in the temporal lobe results in delusions. LBD can occur through two pathways. In the pure form of dementia with Lewy bodies (DLB),

Lewy bodies develop in the brainstem and cortex, and dementia sets in early, leading to DLB. Sometimes, when Lewy bodies develop initially in the brainstem and later expand to the cerebral cortex, with Parkinson's disease (PD) at onset and dementia occurring later (about a year later) in the disease process, it is Parkinson's disease type of LBD or Parkinson's disease with dementia (PDD). Thus, LBD is an umbrella term with two subtypes, i.e., pure form of DLB and Parkinson's disease type of LBD. Lewy body pathology in the brainstem with no dementia leads to Parkinson's disease (Cummings 2004; Outeiro et al. 2019; Haider et al. 2022).

The difference between PDD and DLB is the temporal sequence of appearance of cognitive and motor symptoms. The pure form of DLB and PDD with dementia share many clinical features, with memory disorders being more severe in PDD. The core clinical features of DLB are fluctuating cognition with variation in attention and alertness, recurrent visual hallucinations, rapid eye movement (REM) sleep behaviour disorder, and cardinal features of Parkinsonism like bradykinesia, rigidity, and rest tremors. The supportive clinical features include postural instability, repeated falls, syncope or transient episodes of unconsciousness, hypersomnia, hyposmia, hallucinations, delusions, apathy, anxiety, depression, and autonomic dysfunctions like urinary inconsistencies and hypotension, etc. (Jellinger 2018; Outeiro et al. 2019).

Various factors play a role in increasing the susceptibility of developing LBD. Genes play a role in DLB. Variation in APOE4 and SNCA (α-syn genes) gene is associated with DLB. Glucocerebrosidase mutation and SNCA are associated with both DLB and PDD (Jellinger 2018). Having a family member with DLB increases the risk for developing DLB. Siblings of affected individuals are at more than twice the risk (Outeiro et al. 2019). The other risk factors are age, gender, anxiety, depression, stroke, a family history of PD, and carrying APOE4 genes in comparison to normal counterparts (Boot et al. 2013). Head injury, lesser use of antioxidants, physical inactivity, and diet poor in vitamin E are also reported risk factors (Mrak and Griffin 2007).

Studies have been conducted to estimate the prevalence and incidence rate of DLB. The incidence rates for DLB ranged from 0.5 to 1.6 per 1000 person-years. DLB accounts for 3.2–7.1% of all dementia cases in the incidence studies and 0.3–24.4% in the prevalence studies (Hogan et al. 2016). DLB accounts for 5% of all dementia cases and PDD affects 75–95% of PD subjects. The incidence of both has been estimated to be 6% (Jellinger 2018). Prevalence estimates of DLB range from 0 to 5% in the general population and 0–30.5% of all dementia cases, whereas incidence rates are 0.1% in the general population and 3% of all dementia cases (Galvin 2019). A 30–80% of PD subjects eventually develop dementia. The prevalence of DLB ranges between 2.8 and 30.5% and incidence rates from 31.6 to 112.3 per 100,000 person-years (Fields 2017). It can be seen that the prevalence rate of DLB was as high as 30%.

It has been estimated that 24–31% of PD patients have dementia and 3-4% of dementia cases are due to PDD. The prevalence of PDD in the general population aged 65 and above is 0.2–0.5% (Aarsland et al. 2005).

10.6.4 Frontotemporal Dementia (FTD)

Frontotemporal dementia is a group of complex disorders characterised by loss of nerve cells in the frontal and temporal lobes of the brain. Deposition of abnormal protein aggregates in the frontal and temporal lobes damages the nerve cells and affects their normal functioning. These proteins are microtubule-associated protein, tau; transactive response DNA-binding protein with an Mw of 43 kDa known as TDP-43; and protein fused in sarcoma known as FUS. The abnormal aggregation of TDP-43 protein accounts for almost 50% of FTD cases, whereas tau protein accounts for 45% of cases and FUS 5–10% of cases (Khan and De Jesus 2022; Ghoshal and Cairns 2013).

Clinically, FTD is classified into two subtypes: behavioural variant FTD (BvFTD) and language variant FTD. Language FTD is further divided into two types: semantic dementia (SD) and progressive non-fluent aphasia (PNFA). A marked gender difference is seen, wherein BvFTD and SD are more widespread in men and PNFA is more common in women (Ghoshal and Cairns 2013; Onyike and Diehl-Schmid 2013; Khan and De Jesus 2022). BvFTD is characterised by a change in personality, disinhibition, loss of empathy, apathy, impaired judgement and social habits, changes in conduct and temperament, stereotypic behaviour, changes in eating preference, and decline in executive functioning (Balachandran et al. 2021; Piguet and Hodges 2013). SD is a language disorder characterised by fluency in speech but loss in word meaning, difficulty retrieving names, and the use of less precise terms, followed by memory changes with progression in severity. Progressive non-fluent aphasia is characterised by non-fluent and hesitant speech with speech sound and articulatory errors and agrammatical speech and writing, but the word meaning is retained (Warren et al. 2013; Ghoshal and Cairns 2013).

Among the aetiological factors, genetics plays a very important role in the development of FTD. It has been reported that 40–50% of cases are familial in origin. Mutations in more than 20 genes may be involved in the development of FTD, with chromosome 9 open reading frame (72C9orF72) gene, microtubule-associated protein tau gene (MAPT) gene, and progranulin (PGRN) gene as the most commonly mutated genes, accounting for 50–60% of familial cases of FTD. The other non-genetic factors identified were head injury and thyroid disease (Onyike and Diehl-Schmid 2013; Khan and De Jesus 2022).

It is the second most common cause of the onset of dementia at an early age after AD with a mean age of 58 years. The incidence increases with an increase in age, from 2.2/100,000 at 40–49 years to 8.8/100000 at 60–69 years. The overall prevalence is estimated to be 15–22 persons per 100,000 population (Khan and De Jesus 2022) ranging from 2.0 to 18 per 100,000 (Onyike and Diehl-Schmid 2013).

10.7 TREATMENT AND MANAGEMENT

The treatment and management modalities can be broadly classified into two: pharmacological and non-pharmacological therapies as shown in Figure 10.3. Non-pharmacological therapies are considered as the first line of treatment in neurocognitive disorders, as they are more effective with lesser side effects compared to pharmacological treatment.

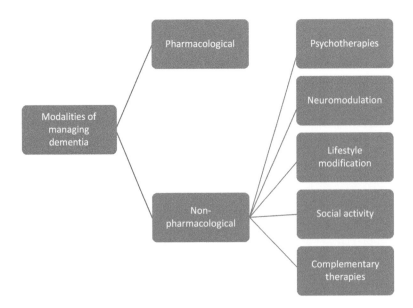

FIGURE 10.3 Management of dementia.

10.7.1 Pharmacological Treatment

Cholinesterase inhibitors, memantine, selective serotonin reuptake inhibitors, and antipsychotic drugs are used in the treatment of dementia and its symptoms (Mathys 2018). Acetylcholinesterase inhibitors and N-methyl-D-aspartate receptor antagonist memantine have been approved for AD depending on the severity of the disease (Apostolova 2016). Vitamin E, in combination with drugs, is also seen to be effective in slowing down the functional decline. These drugs function via multiple mechanisms, such as increasing the cholinergic transmission between the neurones, having a protective effect on neurones, and antioxidative property (Epperly et al. 2017).

Besides, certain drugs can decrease other risk factors leading to cognitive impairment and dementia. For instance, drugs used for treating vascular risk factors, mental disorders like stress, anxiety, and depression may attenuate the progression of cognitive decline. Statins are associated with a decreased risk of dementia (Andrade and Radhakrishnan 2009). Antidepressants are seen to promote neurogenesis, decrease Aβ aggregates, tau pathology, and posses anti-inflammatory properties, thereby decreasing the risk of dementia (Dafsari and Jessen 2020). Similarly, anxiolytic agents can reduce the risk of developing cognitive impairment and AD, especially in E4 carriers with anxiety (Burke et al. 2017). However, owing to the side effects, cost, and often poor adherence to pharmacological treatment, non-pharmacological modalities are preferred.

10.7.2 Non-pharmacological Treatment

Non-pharmacological treatment is considered as the first-line treatment of dementia. It has been found to be effective in treating behavioural and psychological symptoms associated with dementia. It can be classified into two categories: the first is direct intervention where changes in lifestyle are made and psychotherapy is imparted, and the is indirect intervention where modifications in the environment are made and the caregivers are educated to manage and control the symptoms of dementia (Magierski et al. 2020). The various non-pharmacotherapies used in treating dementia and the behavioural and psychological symptoms associated with it are discussed below.

10.7.2.1 Psychotherapy

Certain psychological interventions have gained momentum in treating and managing dementia and mild cognitive impairment. These interventions are helpful in alleviating the symptoms and improving the quality of life of the patient and their caregivers. A few of the psychological therapies effective in cognitive impairment are as follows.

10.7.2.1.1 Reminiscence Therapy (RT)

Reminiscence therapy deals with recalling past events, thereby restoring cognitive abilities, increasing a sense of competence and self-esteem, providing a sense of fulfilment and comfort (Klever 2013; Berg-Weger and Stewart 2017). It is an effective tool in reducing depression, increasing social interaction (Moon and Park 2020), improving mood and cognitive abilities (Cotelli et al. 2012), general behaviour functions, and reducing the strain on caregivers (Woods et al. 2005), and also improving the quality of life and communication (Woods et al. 2018). Sensorial reminiscence has been found to be effective in decreasing behavioural problems and cognitive impairments and increasing functional activities (Deponte and Missan 2007).

10.7.2.1.2 Validation Therapy (VT)

Validation therapy focuses on validating the patient's emotions and feelings, thereby reducing stress and behavioural problems and promoting contentment (Berg-Weger and Stewart 2017). It aims at accepting and communicating the patient's identity/reality, which prevents the expression of agitation behaviour on the patient's part. VT has been found to be effective in decreasing behavioural problems (Deponte and Missan 2007), depression, and improving quality of life (Ra and Lee 2014).

It helps facilitators to relate and communicate with the disoriented elderly and increase motivation and satisfaction among care professionals (Sanchez-Martinez et al. 2021) and therefore this technique can be extended to caregivers as well, so that they do not stress out.

10.7.2.1.3 Reality Orientation (RO)

Reality orientation deals with orienting the subjects with time and place with the aid of puzzles, calendars, games, and reality orientation boards. This helps to decrease confusion and behavioural symptoms (Berg-Weger and Stewart 2017). It has a beneficial effect on cognition and behaviour (Spector et al. 2007), however, contradictory results on depression and behavioural symptoms have been seen by Chiu et al. (2018) but they are in agreement with improvement in cognitive functioning.

10.7.2.1.4 Cognitive Stimulation Therapy (CST)

It is a cost-effective method to improve cognitive and social functioning with the help of cognitive-based activity and tasks (Berg-Weger and Stewart 2017). CST has been seen to improve cognition and quality of life of patients (Spector et al. 2008). CST enhances language comprehension of syntax which may improve communication and reduce depression, thereby improving the quality of life of dementia sufferers which indirectly improves the caregivers' wellbeing too (Toh et al. 2016).

Cognitive training increases speed of processing, reasoning, memory (Andrade and Radhakrishnan 2009), enhances cognitive function, slows the decline of cognitive functions, and reduces the disability arising from cognitive decline (Gates et al. 2011). Frequent, regular, multidomain cognitive exercises have been found to be more beneficial (Gates et al. 2011). However, it has been seen that cognitive training is effective in improving cognitive performance in healthy subjects and not in MCI subjects (Butler et al. 2018) but when cognitive training is imparted along with physical exercise and lecture, it results in improved cognitive and physical functioning (Kouzuki et al. 2020). An intellectually stimulating environment and mental exercise in early life help to maintain cognitive reserves in later life and have a protective effect against cognitive decline (Savica and Petersen 2011) and that is why education has a protective effect on dementia.

10.7.2.2 Neuromodulatory Therapy

Besides the above-discussed verbal therapies, neuromodulation is another effective therapy used for treating dementia and its associated symptoms. Neuromodulation therapy acts directly on the nerves by delivering electrical or pharmaceutical agents directly to the target area in the brain.

Non-invasive and invasive brain stimulation, such as repetitive transcranial magnetic stimulation, transcranial direct current stimulation, transcranial electromagnetic treatment, vagal nerve stimulation, and deep brain stimulation, have been found to have potential in improving cognition (Nardone et al. 2015; Azmi 2020). These approaches are dealt with in detail in Chapter 9. Neuromodulation therapy slows functional, cognitive, and memory decline. It reduces the loss of brain volume, thereby slowing disease progression (Ray 2021).

Photobiomodulation is another non-invasive therapy that is seen to improve cognitive scores. It is a type of light therapy that uses red or near-infrared light to heal and protect dying, degenerating, or injured tissues and provides relief from pain and inflammation. The therapy is suggested to reduce oxidative stress, improve blood flow, tissue metabolism, synaptogenesis, and clearance of Aβ, which consequently has a positive effect on memory, along with a reduction in anxiety, outburst behaviour and agitation in subjects. This consequently reduces the stress of caregivers as well (Hamblin 2019; Azmi 2020). However, more clinical trials are still awaited to establish the role of light therapy in treating dementia.

10.7.2.3 Lifestyle Modifications

Modification in lifestyle can prove to be very effective in not only managing but also preventing cognitive impairment and dementia. Maintaining a healthy lifestyle in synergy with psychotherapies and neuromodulation can be another measure to manage dementia.

10.7.2.3.1 Dietary Measures

Certain nutrients in the form of diet or supplements play a role in preventing or slowing the progression of neurocognitive disorders. Regular intake of fatty fish, fruit and vegetables, and tea is associated with reducing the risk of all-cause dementia (Andrade and Radhakrishnan 2009). Antioxidants such as vitamins E, C, and selenium may be beneficial in decreasing oxidative stress and thus reducing the chances of developing neurocognitive disordersdisorders. as well as slowing the progress of degeneration; however, there is a need to study the dosage and form of antioxidants before administering them in the diet of dementia sufferers (Swaminathan and Jicha 2014).

Western diet has been reported to increase the decline of cognitive abilities, while on the other hand, the Mediterranean diet and Dietary Approaches to Stop Hypertension (DASH) diet have been seen to be associated with better cognitive functioning. Therefore, taking into consideration the beneficial effects of these diets on cognitive functioning, a hybrid diet known as the Mediterranean-DASH Intervention for Neurodegenerative Delay (MIND) diet has been designed by modifying the Mediterranean and DASH diets. In the MIND diet, the components and frequency of consumption of certain foods have been modified. The MIND diet has been found to have a protective effect against cognitive decline and delays the occurrence of neurocognitive disorder (Morris 2016). Adherence to the MIND diet has been suggested to reduce the risk of developing early-onset dementia (Filippini et al. 2020). Consumption of high amounts of cereals, dairy products, ice cream, cake, and coffee increases the risk of developing early-onset dementia whereas leafy vegetables, citrus fruits nuts, chocolate, and moderate consumption of coffee have been inversely related to the risk of early-onset dementia risk (Filippini et al. 2020). Diet rich in soybeans, vegetables, milk and dairy products and low in rice reduces the risk of dementia (Singh et al. 2019).

For the dietary management of dementia, emphasis should be given to modify the diet in such a way that it reduces the biological risk factors of dementia, like obesity, hypercholesterolaemia, diabetes, and hypertension, etc. (Singh et al. 2019). Screening of the nutritional state of the elderly suffering from dementia is recommended as there are chances of malnutrition because of their fragile mental and physical condition. It is suggested that proper nutrition care and support should be an integral part of the dementia management process (Volkert et al. 2015).

10.7.2.3.2 Herbs

Herbs have been part of traditional medicine systems (Chinese and Indian) and Indian cuisine. The bioactive components present in herbs possess various therapeutic and functional properties. A few herbs are known to have neuroprotective, memory-boosting, and cognitive function-enhancing properties.

Gingko biloba, Huperzia serrata, Curcuma longa, Ginseng, Bacopa monnieri, Crocus sativus, Camellia sinensis, Withania somnifera, Uncaria tomentosa, Centella asiatica, Hericium erinaceus, Convolvulus pluricaulis, Emblica officinalsa, and *Semencarpus anacardium* are the herbs possessing neuroprotective abilities. These herbs are seen to be effective in counteracting dementia and its symptoms, improving learning, memory, and cognitive functioning (Chang et al. 2016; Thakur et al. 2018; Gregory et al. 2021). These herbs can prove to be cost-effective and non-toxic therapeutic agents for treating and preventing dementia. Nevertheless, more clinical trials on dementia patients are warranted to establish their dosage and preparation as a single or as in combination formulation.

10.7.2.3.3 Physical Activity

Physical activity has been shown to slow down dementia. Aerobic exercise has a preventing effect on dementia by facilitating neuroprotective neurotrophic factors and neuroplasticity. Besides, it also reduces vascular risk factors contributing to dementia (Ahlskog et al. 2011). Exercise decreases $A\beta$ plaques and neurofibrillary tangles (NFTs), inflammation, and increases brain-derived neurotrophic factor (BDNF) levels and cerebral blood flow (De la Rosa et al. 2020). The risk of dementia decreases in individuals who are engaged in some kind of physical activity. Exercise facilitates neuroplasticity, improves synaptogenesis, hippocampal functioning, cognition, and depression (Andrade and

Radhakrishnan 2009). Therefore, moderate-intensity physical exercise is suggested for lowering cognitive risks and slowing cognitive decline (Ahlskog et al. 2011), yet more research is needed to optimise the type, duration, and intensity of exercises for treating and preventing dementia.

10.7.2.3.4 Meditation

Meditation techniques can be adopted for managing cognitive decline, which may lead to dementia. *Kirtan kriya*, a meditation technique, has been seen to improve sleep, decrease depression and anxiety, downregulate inflammatory genes, and up-regulate immune system genes (Khalsa 2015). Meditation increases cortical thickness and grey matter volume in certain areas related to executive control and memory (Dwivedi et al. 2021).

10.7.2.3.5 Sleep

Proper sleep is important as insomnia, disruption in circadian rhythm, and sleep duration are associated with dementia. Sleeping for less than five hours or more than ten hours was identified as risk factors for dementia (Singh et al. 2019). Insomnia is associated with inflammation, apnoea with cerebral hypoxia and hypometabolism leading to loss of regional cortex and white matter hyperintensities, and longer or shorter sleep duration with faster atrophy, ventricular enlargement, and hippocampal degeneration (Xu et al. 2020). Thus, sleep management is important to prevent cognitive decline and prevent dementia. Bright light therapy, music therapy, physical activity, sleep hygiene education, and certain pharmacological agents could be adopted for management of sleep in dementia (Kinnunen et al. 2017).

10.7.2.4 Social Activity

Social activity refers to maintaining harmonious relationships with other individuals and engaging oneself in social activities. A decrease in social engagements increases the risk of developing dementia (Savica and Petersen 2011). Social engagements increase cognitive reserves and improve cognitive performance (Middleton and Yaffe 2009). High social engagements have been seen to be associated with larger brain volume and grey matter volume and help preserve brain tissues (James et al. 2012). Socially isolated and depressed subjects are more susceptible to dementia, and therefore, dementia prevention strategies should target social isolation. Patients should be encouraged to participate in social activities of their preference and be in touch with family, friends, and relatives.

10.7.2.5 Other Complementary Therapies

10.7.2.5.1 Aromatherapy

Aromatherapy has been found to be effective in improving cognitive functions in AD subjects (Jimbo et al. 2009). Aromatherapy helps to improve sleep disorders, agitation, disturbed behaviour, and quality of life. It has a positive effect on caregivers as it has been seen to reduce the caregiver's stress and burden. It decreases resistance to care in patients and increases motivational behaviour (Holmes and Ballard 2004; Li et al. 2021). Thus aromatherapy as an inhalation approach is effective in managing the behavioural and psychological symptoms of dementia (Li et al. 2021).

10.7.2.5.2 Music Therapy

Music therapy has been seen to improve the cognitive functioning and depression in dementia patients (Moreno-Morales et al. 2020). Music therapy treats agitation and anxiety in dementia subjects (Thakur et al. 2018).

Certain preventive measures like abstaining from alcohol, tobacco, reducing exposure to pollution, maintaining general health like blood pressure, blood sugar, and body weight, increasing qualitative social contacts, and preserving mental health can be adopted. This can be achieved by taking care of holistic health and wellbeing and taking all the preventive measures necessary to maintain optimal physical, mental, and social health. A multidisciplinary preventive approach consisting of cognitive stimulating activities, mindfulness therapy, modifications in diet, physical activity, socialisation, and psychoeducation is effective in improving cognition and sleep; getting relief from shame and anxiety; and maintaining focus and balance (Mehl-Madrona and Mainguy 2017). A healthy lifestyle comprising moderate to vigorous exercise, no smoking, the MIND diet, light-to-moderate alcohol consumption, and engagement in late-life cognitive activities reduces the risk of dementia. The higher the healthy lifestyle, the lower is dementia risk (Dhana et al. 2020).

10.8   CONCLUSION

Dementia is a group of diseases affecting various cognitive domains, especially in the elderly. It is used synonymously with major neurocognitive disorders and includes changes in memory, thinking, judgement, language, attention/concentration, reasoning, perception, and a person's ability to perform daily functions. The number of people with dementia is increasing at an alarming rate, especially with advancing age and in developing countries, thereby posing a burden on individuals and their caregivers in terms of death and disability. Dementia may occur as a result of certain illnesses like Huntington's disease, HIV infection, prion disease, head injury, or cognitive impairment. This may lead to certain disease-related dementias like AD, VD, FTD, and LBD. Several modifiable and non-modifiable risk factors are associated with the risk of dementia. The management of dementia is a major challenge as there is no effective treatment to slow the progression of cognitive decline. However, controlling the modifiable risk factors and promoting the protective factors can be the key to managing dementia. A thoughtful combination of pharmacotherapy, psychotherapy, and lifestyle modifications is suggested to maintain high cognitive reserves in dementia subjects.

REFERENCES

Aarsland, D., J. Zaccai, and C. Brayne. 2005. A systematic review of prevalence studies of dementia in Parkinson's disease. *Movement Disorders* 20, no. 10: 1255–1263. https://doi.org/10.1002/mds.20527. PMID: 16041803.

Ahlskog, J.E., Y.E. Geda, N.R. Graff-Radford, and R.C. Petersen. 2011. Physical exercise as a preventive or disease-modifying treatment of dementia and brain aging. *Mayo Clinic Proceedings* 86, no. 9: 876–884. https://doi.org/10.4065/mcp.2011.0252

Akhter, F., A. Persaud, Y. Zaokari, Z. Zhao, and D. Zhu. 2021. Vascular dementia and underlying sex differences. *Frontiers in Aging Neuroscience* 13: 720715. https://doi.org/10.3389/fnagi.2021.720715

Al-Adawi, S., N. Braidy, M. Essa, et al. 2014. Cognitive profiles in patients with multi-infarct dementia: An Omani study. *Dementia and Geriatric Cognitive Disorders Extra* 4, no. 2: 271–282.

Alzheimer Association Report. 2020. Alzheimer's disease facts and figures. *Alzheimer's and Dementia* 16: 391–460. https://doi.org/10.1002/alz.12068

Alzheimer's Association. 2021. Global dementia cases forecasted to triple by 2050: New analysis shows a decrease in prevalence due to education countered by increase due to heart health risk factors. *ScienceDaily*. www.sciencedaily.com/releases/2021/07/210727171713.htm (accessed 17 June 2022).

Andrade, C., and R. Radhakrishnan. 2009. The prevention and treatment of cognitive decline and dementia: An overview of recent research on experimental treatments. *Indian Journal of Psychiatry* 51, no. 1: 12–25. https://doi.org/10.4103/0019-5545.44900

Anstey, K.J., R. Peters, M.E. Mortby, et al. 2021. Association of sex differences in dementia risk factors with sex differences in memory decline in a population-based cohort spanning 20–76 years. *Scientific Reports* 11, no. 1: 7710. https://doi.org/10.1038/s41598-021-86397-7

Apostolova, L.G. 2016. Alzheimer disease. *Continuum* 22, no. 2: 419–434. https://doi.org/10.1212/CON .0000000000000307

Arvanitakis, Z., R.C. Shah, and D.A. Bennett. 2019. Diagnosis and management of dementia: Review. *JAMA* 322, no. 16: 1589–1599. https://doi.org/10.1001/jama.2019.4782

Ashby-Mitchell, K., R. Burns, and K.J. Anstey. 2018. The proportion of dementia attributable to common modifiable lifestyle factors in Barbados. *Revista Panamericana de Salud Publica* 42: e17. https://doi.org/10.26633/RPSP.2018.17

Azmi, H. 2020. Neuromodulation for cognitive disorders: In search of lazarus? *Neurology India* 68: S288–S296. https://doi.org/10.4103/0028-3886.302469

Balachandran, S., E.L. Matlock, M.L. Conroy, and C.E. Lane. 2021. Behavioral variant frontotemporal dementia: Diagnosis and treatment interventions. *Current Geriatrics Reports* 10: 101–107. https://doi.org/10.1007/s13670-021-00360-y

Berg-Weger, M., and D.B. Stewart. 2017. Non-pharmacologic interventions for persons with dementia. *Missouri Medicine* 114, no. 2: 116–119.

Bir, S.C., M.W. Khan, V. Javalkar, E.G. Toledo, and R.E. Kelley. 2021. Emerging concepts in vascular dementia: A review. *Journal of Stroke and Cerebrovascular Diseases* 30, no. 8: 105864. https://doi.org/10.1016/j.jstrokecerebrovasdis.2021.105864

Boot, B.P., C.F. Orr, J.E. Ahlskog, et al. 2013. Risk factors for dementia with Lewy bodies: A case-control study. *Neurology* 81, no. 9: 833–840. https://doi.org/10.1212/WNL.0b013e3182a2cbd1

Burke, S.L., J. O'Driscoll, A. Alcide, and T. Li. 2017. Moderating risk of Alzheimer's disease through the use of anxiolytic agents. *International Journal of Geriatric Psychiatry* 32 no. 12: 1312–1321. https://doi.org/10.1002/gps.4614

Butler, M., E. McCreedy, and V.A. Nelson. 2018. Does cognitive training prevent cognitive decline?: A systematic review. *Annals of Internal Medicine* 168, no. 1: 63–68. https://doi.org/10.7326/M17-1531

Cadar, D.C., C. Lassale, H. Davies, D.J. Llewellyn, G.D. Batty, and A. Steptoe. 2018. Individual and area-based socioeconomic factors associated with dementia incidence in England: Evidence from a 12-year follow-up in the English longitudinal study of ageing. *JAMA Psychiatry* 75, no. 7: 723–732. https://doi.org/10.1001/jamapsychiatry.2018.1012

Cerejeira, J., L. Lagarto, and E.B. Mukaetova-Ladinska. 2012. Behavioral and psychological symptoms of dementia. *Frontiers in Neurology* 7: 73. https://doi.org/10.3389/fneur.2012.00073

Chai, B., F. Gao, R. Wu, et al. 2019. Vitamin D deficiency as a risk factor for dementia and Alzheimer's disease: An updated meta-analysis. *BMC Neurology* 19, no. 1: 284. https://doi.org/10.1186/s12883-019-1500-6

Chang, D.H.-T., J. Liu, K. Bilinski, et al. 2016. Herbal medicine for the treatment of vascular dementia: An overview of scientific evidence. *Evidence-Based Complementary and Alternative Medicine* 2016. https://doi.org/10.1155/2016/7293626

Chen, Y., C. Lv, X. Li, et al. 2019. The positive impacts of early-life education on cognition, leisure activity, and brain structure in healthy aging. *Aging* 11 no. 14: 4923–4942. https://doi.org/10.18632/aging.102088

Chiu, H.Y., P.Y. Chen, Y.T. Chen, and H.C. Huang. 2018. Reality orientation therapy benefits cognition in older people with dementia: A meta-analysis. *International Journal of Nursing Studies* 86: 20–28. https://doi.org/10.1016/j.ijnurstu.2018.06.008

Choi, J.C. 2010. Cerebral autosomal dominant arteriopathy with subcortical infarcts and leukoencephalopathy: A genetic cause of cerebral small vessel disease. *Journal of Clinical Neurology* 6, no. 1: 1–9. https://doi.org/10.3988/jcn.2010.6.1.1

Cotelli, M., R. Manenti, and O. Zanetti. 2012. Reminiscence therapy in dementia: A review. *Maturitas* 72, no. 3: 203–205. https://doi.org/10.1016/j.maturitas.2012.04.008

Cummings, J.L. 2004. Dementia with lewy bodies: Molecular pathogenesis and implications for classification. *Journal of Geriatric Psychiatry and Neurology* 17, no. 3: 112–119. https://doi.org/10.1177/0891988704267473

Dafsari, F.S., and F. Jessen. 2020. Depression-an underrecognized target for prevention of dementia in Alzheimer's disease. *Translational Psychiatry* 10, no. 1: 160. https://doi.org/10.1038/s41398-020-0839-1

Darwish, H., N. Farran, S. Assaad, and M. Chaaya. 2018. Cognitive reserve factors in a developing country: Education and occupational attainment lower the risk of dementia in a sample of Lebanese older adults. *Frontiers in Aging Neuroscience* 10: 277. https://doi.org/10.3389/fnagi.2018.00277

De la Rosa, A., G. Olaso-Gonzalez, C. Arc-Chagnaud, et al. 2020. Physical exercise in the prevention and treatment of Alzheimer's disease. *Journal of Sport and Health Science* 9, no. 5: 394–404. https://doi.org/10.1016/j.jshs.2020.01.004

Deckers, K., D. Cadar, M.P.J van Boxtel, F.R.J. Verhey, A. Steptoe, and S. Kohler. 2019. Modifiable risk factors explain socioeconomic inequalities in dementia risk: Evidence from a population-based prospective cohort study. *Journal of Alzheimer's Disease* 71, no. 2: 549–557. https://doi.org/10.3233/JAD-190541

Deponte, A., and R. Missan. 2007. Effectiveness of validation therapy (VT) in group: Preliminary results. *Archives of Gerontology and Geriatrics* 44, no. 2: 113–117. https://doi.org/10.1016/j.archger.2006.04 .001

Dhana, K., D.A. Evans, K.B. Rajan, D.A. Bennett, and M.C. Morris. 2020. Healthy lifestyle and the risk of Alzheimer dementia: Findings from 2 longitudinal studies. *Neurology* 95, no. 4: e374–e383. https://doi .org/10.1212/WNL.0000000000009816

Di Marco, L.Y., A. Marzo, and M. Munoz-Ruiz. 2014. Modifiable lifestyle factors in dementia: A systematic review of longitudinal observational cohort studies. *Journal of Alzheimer's Disease* 42, no. 1: 119–135. https://doi.org/10.3233/JAD-132225

Dwivedi, M., N. Dubey, A.J. Pansari, et al. 2021. Effects of meditation on structural changes of the brain in patients with mild cognitive impairment or Alzheimer's disease dementia. *Frontiers in Human Neuroscience* 15: 728993. https://doi.org/10.3389/fnhum.2021.728993

Epperly, T., M.A. Dunay, and J.L. Boice. 2017. Alzheimer disease: Pharmacologic and nonpharmacologic therapies for cognitive and functional symptoms. *American Family Physician* 295, no. 12: 771–778.

Fan, L., C. Mao, X. Hu, et al. 2020. New insights into the pathogenesis of Alzheimer's Disease. *Frontiers in Neurology* 10: 1312. https://doi.org/10.3389/fneur.2019.01312

Fields, J.A. 2017. Cognitive and neuropsychiatric features in Parkinson's and Lewy Body Dementias. *Archives of Clinical Neuropsychology* 32, no. 7: 786–801.

Filippini, T., G. Adani, M. Malavolti, et al. 2020. Dietary habits and risk of early-onset dementia in an Italian case-control study. *Nutrients* 12, no. 12: 3682. https://doi.org/10.3390/nu12123682

Galvin, J.E. 2019. Lewy body dementia. *Practical Neurology*: 67–71. https://practicalneurology.com/articles /2019-june/lewy-body-dementia-1

Gates, N.J., P.S. Sachdev, M.A. Fiatarone Singh, and M. Valenzuela. 2011. Cognitive and memory training in adults at risk of dementia: A systematic review. *BMC Geriatrics* 11: 55. https://doi.org/10.1186/1471 -2318-11-55

GBD 2019 Dementia Forecasting Collaborators. 2022. Estimation of the global prevalence of dementia in 2019 and forecasted prevalence in 2050: An analysis for the Global Burden of Disease Study 2019. *Lancet Public Health* 7, no. 2: E105–E125.

Ghoshal, N., and N.J. Cairns. 2013. Unravelling the mysteries of frontotemporal dementia. *Missouri Medicine* 110, no. 5: 411–416.

Gregory, J., Y.V. Vengalasetti, D.E. Bredesen, and R.V. Rao. 2021. Neuroprotective herbs for the management of Alzheimer's Disease. *Biomolecules* 11: 543. https://doi.org/10.3390/biom11040543

Gudala, K., D. Bansal, F. Schifano, and A. Bhansali. 2013. Diabetes mellitus and risk of dementia: A meta-analysis of prospective observational studies. *Journal of Diabetes Investigation* 4, no. 6: 640–650. https://doi.org/10.1111/jdi.12087

Haider, A., B.C. Spurling, and J.C. Sanchez-Manso. 2022. Lewy body dementia. In *StatPearls [Internet]*. StatPearls Publishing, Treasure Island (FL). https://www.ncbi.nlm.nih.gov/books/NBK482441/

Hamblin, M.R. 2019. Photobiomodulation for Alzheimer's Disease: Has the light dawned? *Photonics* 6, no. 3: 77. https://doi.org/10.3390/photonics6030077

Hamid, T.A., S. Krishnaswamy, S.S. Abdullah, and Y.A. Momtaz. 2010. Sociodemographic risk factors and correlates of dementia in older Malaysians. *Dementia and Geriatric Cognitive Disorders* 30, no. 6: 533–539. https://doi.org/10.1159/000321672

Hasselgren, C., L. Dellve, H. Ekbrand, et al. 2018. Socioeconomic status, gender and dementia: The influence of work environment exposures and their interactions with *APOE* ε4. *SSM – Population Health* 5: 171–179. https://doi.org/10.1016/j.ssmph.2018.06.009

Hogan, D., K. Fiest, J. Roberts, et al. 2016. The prevalence and incidence of dementia with lewy bodies: A systematic review. *Canadian Journal of Neurological Sciences* 43, no. S1: S83–S95. https://doi.org/10 .1017/cjn.2016.2

Holmes, C., and C. Ballard. 2004. Aromatherapy in dementia. *Advances in Psychiatric Treatment* 10, no. 4: 296–300. https://doi.org/10.1192/apt.10.4.296

Hugo, J., and M. Ganguli. 2014. Dementia and cognitive impairment: Epidemiology, diagnosis, and treatment. *Clinics in Geriatric Medicine* 30, no. 3: 421–442. https://doi.org/10.1016/j.cger.2014.04.001

Iadecola, C., M. Duering, and V. Hachinski. 2019. Vascular cognitive impairment and dementia: JACC Scientific Expert Panel. *Journal of the American College of Cardiology* 73, no. 25: 3326–3344. https:// doi.org/10.1016/j.jacc.2019.04.034

Iemolo, F., G. Duro, C. Rizzo, L. Castiglia, V. Hachinski, and C. Caruso. 2009. Pathophysiology of vascular dementia. *Immunity and Ageing* 6: 13. https://doi.org/10.1186/1742-4933-6-13

James, B.D., T.A. Glass, B. Caffo, et al. 2012. Association of social engagement with brain volumes assessed by structural MRI. *Journal of Aging Research* 2012: 512714. https://doi.org/10.1155/2012/512714

Jellinger, K.A. 2013. Pathology and pathogenesis of vascular cognitive impairment-a critical update. *Frontiers in Aging Neuroscience* 5: 17. https://doi.org/10.3389/fnagi.2013.00017

Jellinger, K.A. 2018. Dementia with Lewy bodies and Parkinson's disease-dementia: Current perspectives. *International Journal of Neurology and Neurother* 5: 76. https://doi.org/10.23937/2378-3001/1410076

Jimbo, D., Y. Kimura, M. Taniguchi, M. Inoue, and K. Urakami. 2009. Effect of aromatherapy on patients with Alzheimer's disease. *Psychogeriatrics* 9 no. 4: 173–179. https://doi.org/10.1111/j.1479-8301.2009.00299.x

Kalaria, R.N. 2016. Neuropathological diagnosis of vascular cognitive impairment and vascular dementia with implications for Alzheimer's disease. *Acta Neuropathologica* 131, no. 5: 659–685. https://doi.org/10.1007/s00401-016-1571-z

Kalimo, H., Q. Miao, S. Tikka, et al. 2008. CADASIL: The most common hereditary subcortical vascular dementia. *Future Neurology* 3. https://doi.org/10.2217/14796708.3.6.683

Khalsa, D.S. 2015. Stress, meditation, and Alzheimer's disease prevention: Where the evidence stands. *Journal of Alzheimer's Disease* 48, no. 1: 1–12. https://doi.org/10.3233/JAD-142766

Khan, I., and O. De Jesus. 2022. Frontotemporal lobe dementia. [Updated 30 April 2022]. In *StatPearls* [Internet]. StatPearls Publishing, Treasure Island (FL).

Killin, L.O., J.M. Starr, I.J. Shiue, and T.C. Russ. 2016. Environmental risk factors for dementia: A systematic review. *BMC Geriatrics* 16, no. 1: 175. https://doi.org/10.1186/s12877-016-0342-y

Kinnunen, K.M., A. Vikhanova, and G. Livingston. 2017. The management of sleep disorders in dementia: An update. *Current Opinion in Psychiatry* 30, no. 6: 491–497. https://doi.org/10.1097/YCO.0000000000000370

Klever, S. 2013. Reminiscence therapy. *Nursing* 43, no. 4: 36–37. https://doi.org/10.1097/01.NURSE.0000427988.23941.51

Kouzuki, M., T. Kato, K. Wada-Isoe, et al. 2020. A program of exercise, brain training, and lecture to prevent cognitive decline. *Annals of Clinical and Translational Neurology* 7, no. 3: 318–328. https://doi.org/10.1002/acn3.50993

Kroger, E., R. Andel, J. Lindsay, Z. Benounissa, R. Verreault, and D. Laurin. 2008. Is complexity of work associated with risk of dementia? The Canadian study of health and aging. *American Journal of Epidemiology* 167, no. 7: 820–830. https://doi.org/10.1093/aje/kwm382

Leso, V., A. Caturano, I. Vetrani, and I. Iavicoli. 2021. Shift or night shift work and dementia risk: A systematic review. *European Review for Medical and Pharmacological Sciences* 25, no. 1: 222–232. https://doi.org/10.26355/eurrev_202101_24388

Levine, D.A., A.L. Gross, E.M. Briceno, et al. 2021. Sex differences in cognitive decline among US adults. *JAMA Network Open* 4, no. 2: e210169. https://doi.org/10.1001/jamanetworkopen.2021.0169

Li, B.S.Y., C.W.H. Chan, M. Li, I.K.Y. Wong, and Y.H.U. Yu.2021. Effectiveness and safety of aromatherapy in managing behavioral and psychological symptoms of dementia: A mixed-methods systematic review. *Dementia and Geriatric Cognitive Disorders Extra* 11, no. 3: 273–297. https://doi.org/10.1159/000519915

Lourida, I., E. Hannon, T.J. Littlejohns, et al. 2019. Association of lifestyle and genetic risk with incidence of dementia. *JAMA* 322, no. 5: 430–437. https://doi.org/10.1001/jama.2019.9879

Madhvan, A., G. Bajaj, P. Dasson Bajaj, and D.F. D'Souza. 2022. Cognitive abilities among employed and unemployed middle-aged women – A systematic review. *Clinical Epidemiology and Global Health* 15: 101042. https://doi.org/10.1016/j.cegh.2022.101042

Magierski, R., T. Sobow, E. Schwertner, and D. Religa. 2020. Pharmacotherapy of behavioral and psychological symptoms of dementia: State of the art and future progress. *Frontiers in Pharmacology* 11: 1168. https://doi.org/10.3389/fphar.2020.01168

Mathys, M. 2018. Pharmacologic management of behavioral and psychological symptoms of major neurocognitive disorder. *Mental Health Clinician* 8, no. 6: 284–293. https://doi.org/10.9740/mhc.2018.11.284

Mehl-Madrona, L., and B. Mainguy. 2017. Collaborative management of neurocognitive disorders in primary care: Explorations of an attempt at culture change. *Permanente Journal* 21: 16–27. https://doi.org/10.7812/TPP/16-027

Middleton, L.E., and K. Yaffe. 2009. Promising strategies for the prevention of dementia. *Archives of Neurology* 66, no. 10: 1210–1215. https://doi.org/10.1001/archneurol.2009.201

Moon, S., and K. Park. 2020. The effect of digital reminiscence therapy on people with dementia: A pilot randomized controlled trial. *BMC Geriatrics* 20: 166. https://doi.org/10.1186/s12877-020-01563-2

Moreno-Morales, C., R. Calero, P. Moreno-Morales, and C. Pintado. 2020. Music therapy in the treatment of dementia: A systematic review and meta-analysis. *Frontiers in Medicine* 7: 160. https://doi.org/10.3389/fmed.2020.00160

Morris, M.C. 2016. Nutrition and risk of dementia: Overview and methodological issues. *Annals of the New York Academy of Sciences* 1367, no. 1: 31–37. https://doi.org/10.1111/nyas.13047

Mrak, R.E., and W.S. Griffin. 2007. Dementia with Lewy bodies: Definition, diagnosis, and pathogenic relationship to Alzheimer's disease. *Neuropsychiatric Disease and Treatment* 3, no. 5: 619–625.

Nakahori, N., M. Sekine, M. Yamada, T. Tatsuse, H. Kido, and M. Suzuki. 2018. A pathway from low socio-economic status to dementia in Japan: Results from the Toyama dementia survey. *BMC Geriatrics* 18: 102. https://doi.org/10.1186/s12877-018-0791-6

Nardone, R., Y. Holler, and F. Tezzon. 2015. Neurostimulation in Alzheimer's disease: From basic research to clinical applications. *Neurological Sciences* 36, no. 5: 689–700. https://doi.org/10.1007/s10072-015-2120-6

National Institute of Aging. 2021. What is Alzheimer's disease? https://www.nia.nih.gov/health/what-alzheimers-disease

Nazarko, L. 2019. Dementia: Prevalence and pathophysiology. *British Journal of Healthcare Assistants* 13, no. 6: 266–270. https://doi.org/10.12968/bjha.2019.13.6.266

Ong, P.A., F.R. Annisafitrie, N. Purnamasari, et al. 2021. Dementia prevalence, comorbidities, and lifestyle among Jatinangor elders. *Frontiers in Neurology* 12: 643480. https://doi.org/10.3389/fneur.2021.643480

Onyike, C.U., and J. Diehl-Schmid. 2013. The epidemiology of frontotemporal dementia. *International Review of Psychiatry* 25, no. 2: 130–137. https://doi.org/10.3109/09540261.2013.776523

Ott, A., R.P. Stolk, F. van Harskamp, H.A.P. Pols, A. Hofman, and M.M.B. Breteler. 1999. Diabetes mellitus and the risk of dementia: The Rotterdam Study. *Neurology* 53, no. 9: 1937. https://doi.org/10.1212/WNL.53.9.1937

Outeiro, T.F., D.J. Koss, D. Erskine, et al. 2019. Dementia with Lewy bodies: An update and outlook. *Molecular Neurodegeneration* 14, no. 1: 5. https://doi.org/10.1186/s13024-019-0306-8

Pathak, K.K., and E. Mattos. 2021. Dementia and nutrition. In *Meat and Nutrition*, edited by C.L. Ranabhat. IntechOpen, London. https://doi.org/10.5772/intechopen.96233

Paulson, H.L., and I. Igo. 2011. Genetics of dementia. *Seminars in Neurology* 31, no. 5: 449–460. https://doi.org/10.1055/s-0031-1299784

Piguet, O., and J.R. Hodges. 2013. Behavioural-variant frontotemporal dementia: An update. *Dementia and Neuropsychologia* 7, no. 1: 10–18. https://doi.org/10.1590/S1980-57642013DN70100003

PSSRU (Personal Social Services Research Unit). 2017. *The Effect of Midlife Risk Factors on Dementia in Older Age.* Public Health England, London. https://assets.publishing.service.gov.uk

Qiu, C., M. Kivipelto, and E. von Strauss. 2009. Epidemiology of Alzheimer's disease: Occurrence, determinants, and strategies toward intervention. *Dialogues in Clinical Neuroscience* 11, no. 2: 111–128. https://doi.org/10.31887/DCNS.2009.11.2/cqiu

Ra, S.M., and D.Y. Lee. 2014. P2–340: The effectiveness of validation therapy for dementia patients in Korea. *Alzheimer's & Dementia* 10: P603–P603. https://doi.org/10.1016/j.jalz.2014.05.1019

Ray, F. 2021. Neurostimulation slows decline in mild-to-moderate alzheimer's, trial finds. *Alzheimer's News Today*, 18 March 2021. https://alzheimersnewstoday.com/2021/03/18/non-invasive-neurostimulation-slows-functional-cognitive-decline-alzheimers-trial/

Richmond-Rakerd, L.S., S. D'Souza, B.J. Milne, A. Caspi, and T.E. Moffitt. 2022. Longitudinal associations of mental disorders with dementia: 30-year analysis of 1.7 million New Zealand citizens. *JAMA Psychiatry* 79, no. 4: 333–340. https://doi.org/10.1001/jamapsychiatry.2021.4377

Rocca, W.A., M.M. Mielke, P. Vemuri, and V.M. Miller. 2014. Sex and gender differences in the causes of dementia: A narrative review. *Maturitas* 79, no. 2: 196–201. https://doi.org/10.1016/j.maturitas.2014.05.008

Russ, T.C., E. Stamatakis, M. Hamer, J.M. Starr, M. Kivimaki, and G.D. Batty. 2013. Socioeconomic status as a risk factor for dementia death: Individual participant meta-analysis of 86 508 men and women from the UK. *British Journal of Psychiatry* 203, no. 1: 10–17. https://doi.org/10.1192/bjp.bp.112.119479

Russ, T.C., M. Hamer, E. Stamatakis, J.M. Starr, and G.D. Batty. 2011. Psychological distress as a risk factor for dementia death. *Archives of Internal Medicine* 171, no. 20: 1858–1859. https://doi.org/10.1001/archinternmed.2011.521

Sachdev, P.S., D. Blacker, D.G. Blazer, et al. 2014. Classifying neurocognitive disorders: The DSM-5 approach. *Nature Reviews. Neurology* 10, no. 11: 634–642. https://doi.org/10.1038/nrneurol.2014.181

Samuel, L.J., S.L. Szanton, J.L. Wolff, K.A. Ornstein, L.J. Parker, and L.N. Gitlin. 2020. Socioeconomic disparities in six-year incident dementia in a nationally representative cohort of U.S. older adults: An examination of financial resources. *BMC Geriatrics* 20, no. 1: 156. https://doi.org/10.1186/s12877-020 -01553-4

Sanchez-Martinez, I., R. Vilar, J. Irujo, et al. 2021. Effectiveness of the validation method in work satisfaction and motivation of nursing home care professionals: A literature review. *International Journal of Environmental Research and Public Health* 18: 201. https://doi.org/10.3390/ijerph18010201

Savica, R., and R.C. Petersen. 2011. Prevention of dementia. *Psychiatric Clinics of North America* 34, no. 1: 127–145. https://doi.org/10.1016/j.psc.2010.11.006

Series, H., and M. Esiri. 2012. Vascular dementia: A pragmatic review. *Advances in Psychiatric Treatment* 18, no. 5: 372–380. https://doi.org/10.1192/apt.bp.110.008888

Siberski, J. 2012. Dementia and DSM-5: Changes, cost, and confusion. *Today's Geriatric Medicine* 5, no. 6: 12. https://www.todaysgeriatricmedicine.com/archive/110612p12.shtml

Singh, R.B., S. Watanabe, D. Li, et al. 2019. Diet and lifestyle guidelines and desirable levels of risk factors and protective factors for prevention of dementia: A scientific statement from joint symposium of JAAS and APCNS. *Biomedical Journal of Scientific and Technical Research* 17: 12844–12864. https://doi.org /10.26717/BJSTR.2019.17.003006

Spector, A., B.Woods, and M. Orrell. 2008. Cognitive stimulation for the treatment of Alzheimer's disease. *Expert Review of Neurotherapeutics* 8, no. 5: 751–757. https://doi.org/10.1586/14737175.8.5.751

Spector, A., M. Orrell, S. Davies, and B. Woods. 2007. WITHDRAWN: Reality orientation for dementia. *Cochrane Database of Systematic Reviews* 18, no. 3: CD001119. https://doi.org/10.1002/14651858 .CD001119.pub2

Swaminathan, A., and G.A. Jicha. 2014. Nutrition and prevention of Alzheimer's dementia. *Frontiers in Aging Neuroscience* 6: 282. https://doi.org/10.3389/fnagi.2014.00282

Thakur, A.K., P. Kamboj, and K. Goswami. 2018. Pathophysiology and management of Alzheimer's disease: An overview. *Journal of Analytical & Pharmaceutical Research* 9, no. 2: 226–235. https://doi.org/10 .15406/japlr.2018.07.0023

Tiwari, S., V. Atluri, A. Kaushik, A. Yndart, and M. Nair. 2019. Alzheimer's disease: Pathogenesis, diagnostics, and therapeutics. *International Journal of Nanomedicine* 14: 5541–5554. https://doi.org/10.2147/ IJN.S200490

Toh, H.M., S.E. Ghazali, and P. Subramaniam. 2016. The acceptability and usefulness of cognitive stimulation therapy for older adults with dementia: A narrative review. *International Journal of Alzheimer's Disease* 2016: 5131570. https://doi.org/10.1155/2016/5131570

Tori, K., M. Kalligeros, A. Nanda, et al. 2020. Association between dementia and psychiatric disorders in long-term care residents: An observational clinical study. *Medicine* 99, no. 31: e21412. https://doi.org/10 .1097/MD.0000000000021412

Trani, J.F., J. Moodley, M.T.T. Maw, and G.M. Babulal. 2022. Association of multidimensional poverty with dementia in adults aged 50 years or older in South Africa. *JAMA Network Open* 5, no. 3: e224160. https://doi.org/10.1001/jamanetworkopen.2022.4160

van der Flier, W.M., and P. Scheltens. 2005. Epidemiology and risk factors of dementia. *Journal of Neurology, Neurosurgery, and Psychiatry* 76: v2–7. https://doi.org/10.1136/jnnp.2005.082867

Vance, D.E., J.R. Bail, C. Enah, and J. Palmer. 2016. The impact of employment on cognition and cognitive reserves: Implications across diseases and ageing. *Nursing: Research and Reviews* 6: 61–71.

Volkert, D., M. Chourdakis, G. Faxen-Irving, et al. 2015. ESPEN guidelines on nutrition in dementia. *Clinical Nutrition* 34, no. 6: 1052–1073. https://doi.org/10.1016/j.clnu.2015.09.004

Warren, J.D., J.D. Rohrer, and M.N. Rossor. 2013. Clinical review. Frontotemporal dementia. *BMJ* 347: f4827. https://doi.org/10.1136/bmj.f4827

WHO. 2016. *International Statistical Classification of Diseases and Related Health Problems 10th Revision (ICD-10)-WHO Version for 2016, Fifth Edition.* https://icd.who.int/browse10/2016/en#/F00-F09 (accessed 25 July 2022).

WHO. 2021. *Dementia.* https://www.who.int/news-room/fact-sheets/detail/dementia (accessed 25 July 2022).

Wolters, F.J., and M.A. Ikram. 2019. Epidemiology of vascular dementia. *Arteriosclerosis, Thrombosis, and Vascular Biology* 39, no. 8: 1542–1549. https://doi.org/10.1161/ATVBAHA.119.311908

Woods, B., A. Spector, C. Jones, M. Orrell, and S. Davies. 2005. Reminiscence therapy for dementia. *Cochrane Database of Systematic Reviews* 18, no. 2: CD001120. https://doi.org/10.1002/14651858

Woods, B., L. O'Philbin, E.M. Farrell, A.E. Spector, and M. Orrell. 2018. Reminiscence therapy for dementia. *Cochrane Database of Systematic Reviews* 3, no. 3: CD001120. https://doi.org/10.1002/14651858.CD001120.pub3

World Alzheimer Report. 2014. *Dementia and Risk Reduction.* Alzheimer's Disease International, London. www.alzint.org/u/WorldAlzheimerReport2014.pdf

Xu, W., C.C. Tan, J.J. Zou, X.P. Cao, and L. Tan. 2020. Sleep problems and risk of all-cause cognitive decline or dementia: An updated systematic review and meta-analysis. *Journal of Neurology, Neurosurgery, and Psychiatry* 91, no. 3: 236–244. https://doi.org/10.1136/jnnp-2019-321896

Zhao, Y.L., Y. Qu, Y.N. Ou, Y.R. Zhang, L. Tan, and J.T. Yu. 2021. Environmental factors and risks of cognitive impairment and dementia: A systematic review and meta-analysis. *Ageing Research Reviews* 72: 101504. https://doi.org/10.1016/j.arr.2021.101504

Index